Successful Grape Gr
for Eating and Wine-m

A practical gardeners' guide to varieties, [illegible]andry,
harvesting and processing.

Madresfield Court (painting)

Published by
Groundnut Publishing

SUCCESSFUL GRAPE GROWING
FOR EATING AND WINE-MAKING

First published in 1998 by
Groundnut Publishing, Vinces Road, Diss, Norfolk IP22 3HG

Printed in Great Britain by Eye Press Ltd, Diss, Norfolk.

ISBN 0-9527141-1-6

CONTENTS

Foreword

Introduction

Chapter 1 **Fact and Fable, History**

Chapter 2 **Grapes Suitable for Growing in The British Isles**

Chapter 3 **Planting, Pruning, Training and Propagation**

Chapter 4 **Feeding, Mulching and Maintaining Fertility**

Chapter 5 **Pests, Diseases and a Spraying Programme**

Chapter 6 **Harvesting, Preservation and Vinification**

Appendices

1. Monthly Planner
2. Fertiliser Analyses
3. Grape and Wine Extracts – Analyses
4. Varieties – Yield and Quality Potentials
5. Climate Comparisons British Isles/Continent
6. Hydrometer Tables and Use
7. Fermentation Process
8. DIY Press
9. DIY Mill
10. Measuring Acidity
11. Suppliers, Associations, Bibliography
12. Vineyards in Germany Welcoming Visitors

Index

List of Illustrations

FOREWORD

A great many tomes have been written on the ancient lore of producing grapes, generally implying that amateurs cannot! The author, through many years of practical experience has learnt the hard way, making the mistakes which are necessary to gain knowledge worth imparting to others.

He shows that the technicalities surrounding the myths and fables are not as daunting as you may have thought and explains the cultivation of grapes from preparing the ground to harvesting for the table and for the bottle.

This book is, above all else, logical, easily understandable and lends itself readily for practical use.

...Growing vines need not be complicated... Common sense and tender loving care pays the dividend of worthy harvests for years to come.

STEPHEN READ
Read's Nursery, Loddon, Norfolk. June 1998

Introduction

Modern British Isles viticulture began, most probably with Sir Guy Salisbury-Jones' 1952 planting of the Hambledon vineyard.

There are now a great number of successful vineyards and indeed the industry approaches a size which will require an obligatory contribution to the European Union's wine-lake, yet, nearly half a century on, the notion persists that… 'grapes cannot be grown in Britain', encouraged by mainland European interests and no doubt reinforced by bad experiences in growing unsuitable varieties or good varieties using inappropriate and misguided practices… It takes three full years for a vine to achieve a first crop and so it is essential that the reward fits the effort.

Glasshouse varieties will not bear ripe fruit outside, wallgrapes will not succeed in the open vineyard nor will dessert grapes make good wine, but, as long as varieties are grown which are fitted for their intended use and are properly grown in suitable situations, they are as easily grown as is any other British garden fruit.

Any British garden, however small, can produce good grapes… Any British gardener can grow his own luscious grapes for the table… or for the bottle… and it is the aim of this book that they should!

ALAN ROWE
Brockford, Suffolk. June 1998

Dedicated To...

This book is dedicated to the work of:

Colonel and Mrs Gore-Brown, Mr Barrington Brock, Mr E. Hyams, Gillian Pearkes, Major-General Sir G. Salisbury-Jones and Jack Ward... pioneers of modern English viniculture.

And With Sincere Thanks To...

My wife Anne for her tireless proof reading and indexing.

All our friends in Wine-village Senheim/Mosel and most particularly to Winemaster Dieter Schlagkamp, his wife Inge and their sons Andreas and Thomas for our long friendship, their fine wines and their patient instruction in viniculture.

My good friend Graham Sessions for his time and his splendid photography.

Jancis Robinson, the authority for much historical and varietal information.

Stephen Read for 'The Curate's Vinery' and Glasshouse-grape information...

and Arno Schales for permitting the reproduction of his fine grape studies.

CHAPTER ONE

Fact and Fable

The first few million years

Vitis, the vine, has been on earth for millions of years. Fossil remains of **Vitis sezzanensis,** the modern 'European' vine's ancestor, have been found which are ten million years old.

There are some forty or so species which produce edible grapes, but, by far the most important of these is the European and Central Asian **Vitis vinifera** which is responsible for all but about 0.002% of the world's production.

Vitis labrusca, the vine of North Eastern America, is most likely to have been the 'Vineland' grape seen by Lief Ericsson in 1001, and it may also have been the carrier for phylloxera, which virtually wiped out the European vines in the 19th Century. The Victorian European was an avid plant collector and there were no phytosanitary regulations in Europe at that time. By the end of the 19th Century not only European vines but those in South Africa, Australasia and California had been ravaged by phylloxera, a deadly relative of the greenfly. Unlike other American species **V. labrusca** is not resistant to phylloxera and its other dubious quality is that it, its varieties, and most of the progeny resulting from crossing it with other species and their varieties, make wine with a most distinctively "foxy" flavour which is distasteful to most Vinifera drinkers. It and some of its varieties and hybrids produce quite pleasant eating grapes.

Vitis rupestris and **berlandieri** and especially their hybrids have great natural resistance to phylloxera and they are much used as rootstocks for varieties of **V. vinifera**. The vine is essentially native to the Northern Hemisphere between 30 and 50 degrees... 50 degrees North just crosses Lizard Point.

The 'odd' grape is the Scuppernong Muscadine of the Gulf of Mexico (20-30 degrees North). This **Vitis rotundifolia** is the only grape immune to Pierce's Disease which is a scourge, particularly of the Florida grower. The French Huguenots found it growing in abundance when they set up Fort Caroline, in the mid-sixteenth Century.

They produced their first vintage around 1562... the first vintage in America, or at least the first with supporting evidence. It remains the only reliable grape for growing in Florida. It is a good dessert grape but the wine has a bouquet and taste which is most unattractive to all but those with no choice! Florida growers have persisted in introducing many other varieties, in the hope of producing a more palatable wine, only to have them wiped out by Pierce's Disease.

The University of California at Davis believes that more than 8,000 varieties of grape have been named at some time. There are currently some 1,000 or so in commerce. Most mainland European grapes are grafted, but elsewhere many are on their own roots. Grafting doubles the cost of a vine and so it is understandable that growers are tempted to take a chance. Many vineyards, especially in California, have had to be grubbed out and burned because of phylloxera. There have even been a couple of outbreaks in Sussex and Suffolk in recent years.

Chile has never had any problem with phylloxera, nor has Mexico... perhaps due to phylloxera's inability to thrive in loose sandy soil? But it is surprising that growers in countries with a history of the pest persist in using ungrafted stocks!

The other main problems suffered by the grapevine are virus diseases which debilitate the vine into poor yield or death. It is thus most important that the gardener purchases only good stock from reliable sources... after all, the extra cost for gaining high quality will be spread over a lifetime.

The Last Few Thousand Years and 'The European Grape'

No doubt Hunter gatherers, Neanderthals, and Early Man enjoyed grapes and, since fermentation is a completely natural phenomenon, they quite probably enjoyed inebriation too, but the deliberate manufacturing of wine began only a few thousand years ago. Just how many thousands of years ago is a matter for conjecture, but it was most probably a Sumerian or early Persian accomplishment.

There are many folk tales regarding the discovery of wine. All are unlikely. One of my favourites relates to a concubine of King Jhamshid, 'Persia's King Arthur'.

Now King Jhamshid adored grapes. So much so that he had them stored in great clay vessels, so that he could eat them the year round. It was found that the grapes in one pot had 'gone bad' and they were declared poisonous. The concubine, who had suddenly lost favour, decided to end it all by drinking the 'poisonous' grape liquor. Apparently she returned to court so flushed and vivacious that she and King Jhamshid lived happily ever after.

According to the later Greeks, wine was not discovered by man but given by

Dionysius (the Roman Bacchus) the God of Pleasure, to complement mankind's rationality.

Homer (circa 1000 B.C.) described wine as 'part of Society's Fabric'. Vines were cultivated in Egypt before 3000 B.C. for food and drink, at a time when the God Osiris was named "Lord of Food and Wine".

Virgil wrote of Love and Wine as did the somewhat coarser Rabelais. Dante called it bottled sunshine and the Bible has many references in praise and otherwise.

We are told that it was Noah who planted the first vineyard and there is a cautionary Arab story told of him which is not in the Bible...

One year he fertilised his grapes with peacock blood and the grapes tasted well, as did the wine. The next year he used lion blood and he said that the wine was better. In the third year he used monkey blood and thought the wine to be the best ever and in the fourth year he fertilised with pig blood, when he boasted that the wine was unsurpassable.

The warning is, that the drinker's perceptions are not the same as those who, sober, watch his behaviour deteriorate, glass by glass, to that of the trough.

Similar warning comes in a Japanese proverb, "At the first cup you drink the wine. At the second the wine drinks the wine. The third cup of wine drinks you!" And Ecclesiasticus 31 tells us that "Wine was created to make Mankind joyful, not drunken". Horace (1st Century B.C.) wrote, "Bacchus opens the gates of the heart", and, early in the first millennium, Ovid said that "Wine gives courage and make Man apt for passion". Some fifteen hundred or so years later Shakespeare observed, "It provokes desire, but taketh away the performance".

Millions of words have been written and thought of 'The Grape', but of course, Oenology... the study of wines, and Ampelography... the study of vines, cannot take place on the move.

The vines must be tended and there must be a reasonable expectation of 'harvest' and so the essential precursor to the deliberate practice of planting to harvest is, settlement... Civilisation... 'having advanced beyond the primitive savage state; refined in interests and tastes'. The European viniculture seems to have spread westwards from around where we now call Iran. This theory is supported by language: Early Sanskrit, vena; Greek, oenos; Latin, vinum; French, vin; German, Wein, and our own word, for it is not feasible that the cultivation of vines went against the flow of the westwardly spreading 'civilisation'.

The vine spread to Egypt and the Eastern Mediterranean and thence to the Greeks, where in 500 B.C. Herodotus wrote our first known descriptions of winemaking and palmwood coopering. Later viticulture spread to the Romans and elsewhere throughout Europe. The great expansion of viniculture throughout Europe is due, in great measure, to the Romans. They were unconscionably vinolent and set up vineyards wherever they extended their bounds... The Wrotham Pinot ('Dusty Miller') is said to have been brought to Southern England by them.

The first variety to spread westwards was most probably very much like Muscat. The other early emigrants were very similar, if not identical to, Syrah, Pinot Noir and Merlot. The vine of the present day is hermaphrodite and so its propagation does not necessarily need to be sexual... it can produce mutant buds (sports)... The Pinot Noir, for example, has a number of mutants of great value in cultivation, at present.

The exception to this generalisation is that some modern grafting stock is male and, fancifully, called 'Mustang stock'. The species vines are generally dioicous, and the early vinegrower realised that some vines never had fruit (male), some had some fruit sometimes (females with access to varying amounts of pollen) and some were reliable fruiters (hermaphrodite) and these were the best for viticulture and so were repeatedly reproduced vegetatively. Wherever grapes were cultured they would be improved. They would be grown from seed, which would either have been self fertilised or were crossings... in either event differing from the parent(s) or they would be noticed as useful sports.

Newer clones and varieties would move on... The Malvasia, now the grape of Madeira Malmsey, originated in Greece and moved throughout Western Europe before arriving in Madeira two thousand, or more, years later.

Some varieties are transient but others stay in culture for millennia... for example the Chasselas, which is now the standard roadside dessert grape in France and which was much prized by Louis XV as 'Chasselas de Fontainebleau', bears a most striking resemblance to grapes in Egyptian Tomb Paintings of 3500 B.C. and to grapes which are still grown in southern Egypt. The Gutedel of Germany and the Fendant of Switzerland are the same variety.

The Chasselas may have been brought to France by the Phoenicians around 1500 B.C. or around 600 B.C. by the Greeks, when they founded Massilia... now called Marseilles. They had colonised much of the Italian peninsular and Sicily some 200 years earlier and continued the Phoenicians' vine and wine trade all around the Mediterranean. The Muscat of Alexandria (Lexia in Australia) has a self evident source. It and other muscat varieties were the most commonly distributed by the early traders and colonisers, for they are the most luscious of grapes; intensely flavoured and most strongly scented... some etymologists connect muscat, the grape, with 'musca', the fly, which it so compellingly attracts.

Presumably the Musk Ox too, has many attendant flies!

It was the Romans who began French viniculture, most probably with Muscat and nearly 2,000 years on it was the parent of many a Victorian's hothouse hybrid. These days most of our dried grapes come from California – whose boast it is that they alone truly 'sun-dry' their fruit – but in Victorian times and previously, the 'currant' came from Corinth and the 'sultana' from exotic Ottoman Asia Minor.

At the passing of the Romans' Empire, European viniculture became almost exclusively a part of the Christian Church… "the Blood of Christ". The vineyards were maintained for food and for wine, and monasteries became the focus for the breeding of new varieties. It is the great ease with which the vinifera vine will sport and hybridise and the vine's ability to adjust to and thrive in so many provenances, which allow the grape apparently limitless variety and style.

The Roman Catholic Church continued and strengthened the Roman regulatory practices and began, what are with us to this day… wine tax and duty … people will pay dearly for that which they like sufficiently!

The later Norman rulers of England annexed most of France and Henry Plantagenet's marriage with Eleanor of Aquataine brought with it, the principal wine port of Bordeaux and great profit for the Royal purse!

During the reign of Elizabeth, two of the major ailments of the rich were 'Stones' and 'Dropsy'. William Turner, the Queen's physician wrote in 1578 that they were 'caused by Red Bordeaux' and that drinking Rhine white wine would be efficacious.

The Vinifera Leaves Europe

In around 100 B.C. the European grape went East to China and to India and in the 16th Century to Japan, but the first ones to be planted to the west of Europe were in Madeira, originally a very densely forested island… Although known to Arab sailors the first to land there was O Zarco ('The cross-eyed'), in 1419, to claim it for Portugal. It was of little immediate use, since for some reason known only to himself he set the forest on fire and it is recorded that Madeira 'burned as a beacon for sailors for seven years! It may have been his punishment to be made Governor, for he held the post until his death 40 years later.

The Malvasia grape must have arrived soon after Madeira was settled, since the Venetian Da Mosto reported in 1455, that 'the vine plantations of Madeira were the finest sight in the world.' Although Columbus lived on the island around that time, there is no suggestion that he took grapes to the Indies. It was Cortez who introduced the European grape to America in 1524 and by the end of the century the Conquistadores had vines planted from Mexico to Chile. The varieties would eventually include new varieties grown from seeds from the grapes taken by the settling priests. (Raisin... Latin 'a bunch of grapes'.)

Grape growing began in California in 1697 with the introduction of a Chilean variety, by the Jesuit Father Juan Ugarte, to the Mission San Francisco Xavier.

This 'rustic' wine grape, still grown in California as 'Mission', is closely related to the variety Pais which continues to be the principal red winegrape of Chile.

Further plantings came with the different immigrant nationalities: the 18th Century French and Portuguese took native grape varieties to South America; German, Hungarian and French varieties arrived in the U.S.A. during the 19th and 20th Centuries; and by this century too, Italian varieties arrived in Argentina and Californian varieties crossed to its eastern and northern neighbours.

Heugenots and Dutch took German and French grapes to South Africa during the last 50 years of the 17th Century and in 1758 Captain Phillips took these varieties on to Australia. By 1820 they were in New Zealand. Californian varieties went to Australia in the 1950s.

Japan was stocked with French stock in the 1970s and with Australian varieties during the 1980s.

Whilst it had taken more than 6,000 years for **Vitis vinifera** to drift from Trans-Caucasia to Madeira, it exploded around the rest of the globe in only 500 years!

One hundred years ago the European vine was almost lost to us by the ravages of **Phylloxera vastatrix,** imported from North Eastern America along with Powdery Mildew by enthusiastic pathologists and insect collectors in the mid-1800s.

It appears that it was very hard to isolate the cause of the vine's swift collapse, for such was the arrogance of the times that 'authorities' persistently rejected the evidence which stared them in the face... that the threat came from a tiny creature. Many preferred to believe that it was that the vines were degenerating because of repeated vegetative reproduction and that what was needed was... "an injection of 'Male vigour'."

Similarly it was abhorrent to consider that the culprit grew through seventeen stages before maturity, thirteen of which required no male intervention, or that one tiny sexually produced fundatrix female had the potential to result in five hundred million progeny in one season.

Only a few years previously, the 'English Country House' wall grapes had succumbed to Powdery Mildew as had Continental vineyards and a few years earlier the whole of Europe had suffered the Great Potato Blight; which had been said to be 'wholly contained in Belgium' upon its arrival from America. These plagues had similar responses from 'The Authorities' as did phylloxera and as, much more recently, did Dutch Elm Disease, Listeria, B.S.E. and E. coli... for it seems to be obligatory that they observe certain rehearsed stages...

Firstly, to deny the existence of the problem... and then, to deny its importance. Thirdly to deny blame and apportion it elsewhere... and, finally, to take the credit for any resolution.

The opprobrium for phylloxera was correctly given to the U.S.A., but it received scant recognition for supplying the 'cure'... grafting onto other American rootstocks... for, even after more than a hundred years, this remains the only reliable defence against **Phylloxera vastatrix**... 'The leaf drying destroyer'.

As this book is in writing it has been reported by the Dutch Government that Brown Rot (Pseudomonus solanacearum) a bacterial disease of potatoes and potentially a greater threat to the world's potatoes than is blight, has been... "successfully contained in a very small area"...?

Modern Times

Californian experience confirms that phylloxera is alive and well and has evolved new successful strains.

There are presently no phylloxera in Great Britain.

It is a notifiable pest and would result in the destruction of the infested vineyard, without compensation, whether of twenty or twenty thousand vines.

A large proportion of Britain's vines are on their own roots and will succumb if it is introduced, which is reason enough for maintaining a rigorously enforced phytosanitary policy. The grower is in control, in so far as he/she grows grafted stock or maintains isolation and never buys stock from other than reputable sources.

Whilst the United Kingdom is kept free from phylloxera, all the gardener has to attend to is the growing of suitable varieties and growing them well.

And Noah began to be an husbandman and he planted a vineyard.
(Genesis 9, 20)

Chapter One Summary

★ 99.998% of the world's grape production is from V. vinifera varieties.

★ The Sumerians or Early Persians were most likely to have been the first to cultivate the grape.

★ The Phoenecians, the Greeks and the Romans spread viticulture around the Mediterranean basin and thence to the British Isles.

★ Some common varieties seem to have remained unchanged over thousands of years. For example, Chasselas.

★ Grape varieties can evolve sexually and asexually.

★ European vines were almost totally extinguished in the 19th Century by phylloxera, which had been introduced by plant and insect collectors.

★ There is no control for phylloxera, which is notifiable.

★ Vines must be on resistant stocks or in a phylloxera free provenance.

★ The British Isles are free from phylloxera, but only strict Phytosanitary regulation will keep them so.

CHAPTER TWO

Grapes Suitable for Growing in The British Isles

Of the thousand or so vine varieties in commerce, there are about 150 which are likely to be successfully grown in the United Kingdom, either in the open or in unheated greenhouses and polytunnels. There are others which may be grown with heat but few people now can afford to heat grapehouses and so they will be dealt with very briefly at the end of this section. Generally speaking they are Mediterranean varieties or warm country hybrids.

The word 'hybrid', with respect to European Viticulture, means any variety which is not 100% **Vitis vinifera,** for quite obviously all varieties are hybrid in the strictly botanical sense.

The European Community Wine Regulations forbid the use of 'hybrid' grapes for the production of 'Quality Wine' but allow them for 'Table Wine'. The French are the most purist, no doubt remembering that phylloxera was 'foreign'. Their xenophobic attitude is extended to all varieties of **V. vinifera** which are not 'French', by banning the growing of 'non-French' varieties for wine production, more than fifty miles inside the borders. Paradoxically, in most French winegrowing areas there are 'French' grapes only because there are 'Foreign' rootstocks upon which to grow them.

German hybridists are more sanguine, for, whilst they too draw the line at 'foxy' hybrids, much of their effort has been devoted to making **Vitis amurensis** X **V. vinifera** varieties some of which have much to commend them.

In general: hybrid vines crop more heavily; are rather more disease resistant and produce faster maturing wines but they have little capacity for bottle life and are, all in all, inferior to purely **vinifera** varieties.

In choosing this list, I have considered that it is essential for a variety to have these qualities...

1. It must ripen its wood sufficiently to survive our Winter and to have made viable fruit buds for the next year's growth and fruiting, and it must do so well before our first air frosts.

Muscat Hamburg... Veraison

2. It must be likely to set and to ripen a worthwhile crop to the North of its usual provenances.

3. It should have a reputation for disease resistance.

4. It should be able to produce useful pollen in damp, cool and generally less than optimum Spring conditions.

5. It should be of good lineage and likely to reflect the dessert or wine qualities which it shows in its natural home.

6. If it is a winegrape then it must be known to produce drinkable wine in our climate.

Classification

Range

The vines which are listed may be grown successfully and with reasonable certainty, SOUTH OF A LINE FROM THE WASH TO GALWAY. There are commercial vineyards which are to the North and succeed, and certainly the range can be extended to Scotland by selecting early ripening varieties grown in coldhouses or polytunnels, especially if a little warmth can be given in early Spring and the Autumn.

Season

By far the most grown outdoor grape in the United Kingdom is still, Müller-Thurgau, and the grapes which are listed are considered as ripening earlier than, with, or later than Müller-Thurgau.

It is essential for United Kingdom viticulture that a grape ripens its wood sufficiently to survive our Winter. Only those which are noted as ripening their wood at least 'Fairly' should be attempted in the open vineyard.

The various methods of pruning and training are mentioned elsewhere in this book.

Key

Training		Fruit Ripening		Cane Ripening	
Cane pruning	C	Early	E	Good	G
		Mid-Season	M	Fair	F
Spur pruning	S	Late	L	Poor	P

Wherever the parentage is known it is shown.

(Female parent X Male parent)

The most common synonyms follow the classifications.

FRENCH WHITE vinifera varieties

Auxerois C M G

Much grown in Luxembourg and Alsace for making everyday wines and recommended by EU for British culture. Quite vigorous with a potential for good crops. It has been planted commercially in the U.K. Best on early and alkaline sites.

Chardonnay C L G

One of the Great Classic Whites. The grape of Chablis. Champagne and fine Burgundy… but not here! It is capable of producing 'fruity' wines with good balance in a good year and on very early sites or under cover. Chardonnay will thrive only in alkaline conditions. A slow starter but a healthy variety with beautiful golden grapes which, though small are good enough for dessert.

Chasselas C L F Gutedel (Germany), Fendant (Switzerland)

There are many differently coloured clones of this ancient grape, but they are culturally similar. It is a very good dessert grape but produces a flabby wine. The Swiss would dispute this since Fendant is their most grown wine-grape… it may be that it likes a hot mountain behind it! It is well worth growing as a dessert grape under glass or on a south facing wall.

Chenin Blanc S L F Pineau de la Loire

A Loire grape of high quality. Some commercial plantings have been made in the U.K. A good dependable cropper with large wide shouldered bunches. Recommended for protected growing only.

Gewürztraminer C L G

Grown extensively in Alsace, this 'French' grape is a 'German' variety which originated in Italy (Tramino)! It produces a most full bodied fruity wine with a spicy bouquet and taste. If it is Guyot trained it must be highly arched or its sap may miss the first buds. It is better Mosel Hertzliche ('Heart-shaped') trained. It needs a very good site but well worth growing under cover.

Madeleine Angevine C E G

Originally imported by the English vinicultural pioneer Barrington-Brock, this is a consistently good cropper which makes good rich wine with a fine muscat bouquet. It sets its fruit well, even in a poor year. It is the mother of Siegerrebe and the grandmother of Reichensteiner, both of which do well here. It can ripen as early as September. Its wood ripens very well. A British clone is available.

Madeleine Silvaner C E G (Madeleine Royale X Silvaner)

Capable of exquisitely fragrant wines. It is the earliest ripener and is not fussy about site.

Perle de Czaba S E G

A very old variety and best grown on a south wall. It may be used as a dessert grape and it can be made into most acceptably good wine. It is a characteristic of old French varieties that the first fruiting bud is the sixth along the spur.

Perlette S E G

A very old French seedless variety producing good crops of small sweet fruit.

Pinot Gris C M G Rülander (Germ) Tokay d'Alsace

A grey-blue grape. It needs a very good site and can produce wines of high quality and with a most beautiful bouquet. I have heard that it blends well with Auxerrois. At its best bone dry.

Précoce de Malingre S E G

A very reliable cropper, this grape is a very pleasant eating grape when fully ripe, and it makes refreshingly crisp wine. A most suitable grape for late sites.

Sauvignon Blanc C L G

Ubiquitous in France, it may just succeed here on the very earliest of sites. This, the classic grape of Sauternes, is well worth a try under cover.

Gewürtztraminer

Rülander (Pinot Gris)

Müller-Thurgau

Müller-Thurgau (Rivaner)

Ugni Blanc C L P Trebbiano (Italy)

One of the most planted white grapes in France, it is a native of Italy. It produces prodigious crops. It is included here only because in its many guises it is the world's largest single producer for wines and spirits. It is an ancient grape and may be the Trebbularum mentioned by Pliny nearly two thousand years ago. It is far too late for planting outside anywhere in Britain, but it is grown in polytunnels by some English commercial winegrowers. Whilst being a good eating grape it makes totally odourless and featureless wines. Jancis Robinson, in 'Vines, Grapes and Wines' suggests that the risky business of nurturing this vine could be successfully circumvented by mixing alcohol, tartaric acid and water in the laboratory!

GERMAN WHITE vinifera varieties

Müller-Thurgau C M F/P (Riesling X Silvaner)?

This variety is listed out of order because it is the most widely planted white wine variety in this country. It has always been the most highly recommended, but, frankly there is very little to be said in its favour. In fact I wonder if there was a European plot to kill British viniculture at birth.

On anything approaching good soil it produces far too much wood which ripens poorly and leaves it open to frost damage, botrytis and phomopsis. It needs constant attention to the removal of sub-laterals all Summer. It sets poorly in a cold damp year, when it will drop its fruit due to millerandage and coulure.

It can make pleasant 'muscat' wine in a good year, but it can all too often have a 'mousey' or 'catty' nose and taste. This grape is called Rivaner in Luxembourg where it is much grown. It is the most widely grown grape along the Rhine for producing the oceans of Liebfraumilch and other low quality blended wines. Over a quarter of Germany's planting is of Müller-Thurgau and it produces a much bigger fraction of its wine.

It was produced by Dr Müller of Thurgau in 1883 as 'combining the quality of Riesling with the earliness of Silvaner'. Most German growers ignored it, being, quite rightly, dubious of its worth, preferring to continue growing Silvaner for 'everyday' wines and Riesling for 'treats'. It was fifty years before it took hold but after the war its potential of 20,000 litres per hectare in Germany (Riesling 8-10,000) proved to be too great a temptation for all but the finest winemasters. It is unlikely to improve significantly with age and has a life of no more than three years in the bottle. One final indignity is that it is now thought to be no more than a chance Riesling seedling of unknown male parentage.

If I was planting a new vineyard I would not plant a single one!

Bacchusrebe C M F (Silvaner X Riesling) X Müller-Thurgau

Similar to Müller-Thurgau but with better habits and with the potential for making better wine. There are commercial plantings in U.K.

Ehrenfelser C L F/G (Riesling X Silvaner)

A grape of very high quality capable of very high Oechsle levels (sugars are measured as specific gravity... see later) and so capable of producing excellent dessert wines. Too late for outside but worth growing with cover.

Faberrebe C M F (Pinot Blanc X M-T)

Tolerant of soil type, but requiring a good aspect, this grape makes fresh, fruity and racy wines. It does not ripen its wood reliably but could do well under cover.

Findling C E G (M-T X)

Better in every aspect than its parent. It is favoured by some commercial growers in the U.K.

Kanzler C M F (M-T X Silvaner)

A grape of high quality, as 'Chancellor' suggests. It needs a very good site, for it is a tardy wood-ripener. The wine is very fruity and with a splendid nose. Its potentially poor wood makes it uncommercial in Britain, but its capability for making beautiful wines should attract some enthusiasts... especially if grown on a south wall.

Kerner C L G (Trollinger X Riesling)

This is a grape of the highest quality. It is good to eat, it makes beautifully flavoured juice and exquisite wine, especially dessert wines which can be of great elegance. It ripens a little before Silvaner and is so borderline for outdoors even in the South where it is more likely to succeed on a South wall, as a pot-grape or with late Summer cover. Created in 1969 in Wurtemberg its quality approaches that of the superlative Riesling, yet with nuances of its own. A worthwhile challenge for the British gardener. **Kernling** is said to be a mutant of Kerner. It is both earlier ripening and less acidic than its parent. The author has no experience of this variety but if it is what it is said to be, then it must be preferable to **Kerner** in Britain.

Morio-Muscat C L F (Silvaner X Pinot Blanc)

A very heavy cropper but, fortunately, this remarkably blowsy grape is too late for Britain. It is much used to 'spice up' otherwise tasteless 'plonks'.

Bacchus (rebe)

Ehrenfelser

Kerner

Riesling

Ortega C E F (M-T X Riesling)

A modern cross (1971) this grape can attain high sugar levels. It does not like a cold autumn and if very ripe it can be deficient in acid. It is often used for improving the sugar levels in blends. It has very soft bunch stems which requires its being destemmed before winemaking.

Perle C E G (Gewürztraminer X M-T)

This grape is accommodating of site, tough, copes with a poor Spring, is relatively disease free, early, and of moderate cropping potential. It ripens its wood well, is a fair eating grape and produces a light flowery wine. Recommended.

Regner C M G (Lugliencabianca X Gamay Fruh)

It is remarkable that the crossing of an Italian dessert grape with an early clone of a classic French black wine grape should result in a grape which is, not only tolerant of our climate but able to cope with a poor Spring and still set a fair crop of fine fruit! It likes a good site but dislikes chalk.

Reichensteiner S/C E F M-T X (Madeleine Angevine X Calabresser)

This is a grape which performs far better here than it does at home. This is perhaps because it appreciates the long term ripening periods which we so often experience with our 'Indian Summers'. It makes a fair wine with a fine nose.

Riesling C L G

Widely considered by many as the 'first amongst equals', so far as classic white grapes are concerned. It is the last variety to ripen in Germany.

Some of the finest wines in the world are made from the German Riesling but it is much too late for us. It can be grown successfully under glass but its wine will be a pale shadow of German Riesling. Indeed, although the Riesling is so accommodating of provenance, so long as the season is long enough, nowhere else in the world does it perform as elegantly as in Germany 'North of the Bodensee', or Alsace, or as perfectly as in the Mosel, grown on steep slate winehills!

It is definitely a native German grape and most probably a descendant of the wild **Vitis vinifera subspecies sylvestris**. The story goes that it was the Romans who took the 'German' vines off the trees and put them into the 'properly straight rows' of the vineyard!

Scheurebe C L G (Silvaner X Riesling)

An excellent grape, but unfortunately rather late for Britain. It needs shelter from winds when it will hold its grapes well into the Winter. It is capable of very high sugar levels without losing its acid balance and so makes elegant and harmonious wines. Well worth a place in the greenhouse.

Schönburger C M G Pinot Noir X (Chasselas X Muscat Hamburg)

Like Reichensteiner this grape does better in the U.K. than in Germany where it can be mediocre. It can be used as an eating grape and it makes full wines of excellent flavour and bouquet. If I was planting a new vineyard in Suffolk this would be my first choice. It does well on a south wall as a small dessert grape.

Siegerrebe C E G (Madeleine Angevine X Gewürztraminer)

Like the previous grape, an excellent performer in Britain. Although small for dessert, the grapes have a most pleasant spicy taste. The wine is spicy too, and sometimes with a suggestion of grapefruit. Perhaps one year in ten it ripens early enough to suffer wasp damage. It will not tolerate lime. Highly recommended but for this reservation.

Silvaner C L G

Once the workhorse vine of Germany, it is very choosy of site, more so even than Riesling. It dislikes lime. Still sometimes grown under glass in U.K. as a dessert grape. Far too late for Britain out of doors.

Würzer C E G (Gewürztraminer X M-T)

This is a good early grape with fragrance from its mother. Its wine needs to bottle age to develop its pleasant nose and taste. Another good German cross.

HYBRID WHITE GRAPES

Seibel 5279 S M G Aurora (U.S.A.)

Good eating grape with a spicy flavour. The most widely planted grape in the North-Eastern U.S.A.

Seyval C M G

The second most popular wine grape in England. Best on chalky land. Makes drinkable wine but with very little 'length'. It can suffer from fruit dropping and the yield can vary from one half to four kilos per vine according to the Spring weather, but it will have a crop even after a poor Summer. The gardener with aspirations to a little commercialism should remember that being a 'hybrid' its wine and its blend may only be classed as Table Wine in the EU. It is often planted along with M-T and their blend is better than the varietal wines.

GERMAN 'NO SPRAY' HYBRIDS

Perhaps 'Low Spray' would be a better description for these relatively new German hybrid varieties. The parents are unnamed number referenced French hybrids and they are said to have the potential for very high yields without the need for spraying to control mildew or botrytis.

The most accessible for the amateur is Phoenix. It most certainly would be of great advantage to dispense with the need to spray, but reports suggest that whilst Phoenix seems to be resistant to mildew, it can be very susceptible to botrytis. It is my opinion that there may be something of the Philosophers' Stone about 'No spray'... Nothing is for nothing.

FRENCH RED GRAPES

Merlot
Cabernet Sauvignon
Cabernet Franc
Petit Verdot C L G

The classic four of Clarets. Each Chateau has its secret blend but generally the blend is 40-50% Merlot, 40-50% Sauvignon, 5% Franc and a dash of Petit Verdot for nose. All are too late for outside culture, but will ripen in tunnels. Well worth growing, but do not expect Premier Crus unless the greenhouse effect really takes hold, and all importantly, your 'terroir' is ideal.

Gamay S L G

There are many clones available of this, the Beaujolais grape, and it is essential to obtain a northern one, say from the Loire. Being an ancient grape, it produces fruiting buds at the sixth bud and beyond. If you have a really hot sunny site you may succeed with this grape outdoors, but it is more likely to succeed under cover.

Pinot Meunier C M G Wrotham Pinot (Brit) Müllerrebe (Germany)

This vine is said to have been brought to England by the Romans and if this was the case it is our naturalised grape. It is certainly tough enough for our climate,

Scheurebe

Siegerrebe

Silvaner

Schönburger

since it was lost in the wild for hundreds of years before being rediscovered on a Wrotham wall. It buds late. It is often called 'Dusty Miller' because of its very downy leaves. It is a sport of Pinot Noir, to which it is inferior as a wine grape in every respect except that, quite obviously, it tolerates our climate.

Pinot Noir C M G

The noble grape of Burgundy and Champagne. This is the grape which proves the importance of 'terroir', that unique blend of climate, geography and geology, for nowhere outside Burgundy does it produce anything other than vaguely similar wines. Such is the elegance of fine Burgundy that viniculturalists the world over have planted this vine with high hopes and little else. It refuses to perform away from home! It will succeed only on an alkaline site. Just as in Champagne, far to the north of its home, it will produce white wine, whether to be sparkled or not. Some good (for outside Burgundy) Pinot Noirs are grown in the Southern Counties. Generally speaking it results in poor, thin, acid wines when grown outdoors in Britain.

FRENCH HYBRID RED GRAPES

Siebel and Kuhlmann were two of the most successful 19th Century hybridists in breeding American species' phylloxera resistance into European 'nobility'. Time proved that grafting was the solution to the phylloxera problem, but one of Siebel's whites (5219 Aurora) and two of Kuhlmann's reds are still grown widely. Gillian Pearkes, the founder of Yearlstone Vineyard, rated these Kuhlmann Reds as the best of the hybrids.

Leon Millot C E G (Kuhlmann 192/2)

Bears a heavy crop and responds well to Alsace high arched training. A blend of 25% Millot and 75% Triomphe d'Alsace makes a good richly coloured wine which is superior to either varietal.

Triomphe d'Alsace C E G (K 319/3 Knipperle X V. riperia X V. rupestris)

Very vigorous and disease resistant except to coulure and millerandge in a cold season. The varietal wine is very well coloured but it has an extremely powerful taste which needs a couple of years to attenuate. E.C. rules allows only the appellation 'Table Wine' for wines from these 'Hybrid' grapes, but they are reliable croppers in most years in the south and I would expect them to perform well on a south wall at least as far north as the 'Wash-Galway line'.

GERMAN RED GRAPES

Dornfelder C M/L G (Helfensteiner X Heroldrebe)

This grape does well in the south of England where it makes a good light red wine and in addition it is a lovely dessert grape, having large, open and wide shouldered bunches. Further north it appreciates some cover. It has the potential to be a major 'pink' wine grape in the U.K.

Dunkelfelder C E G (Farbertraube IV/4 (4) X Frölich)

Produces a huge crop of tiny grapes with deep red juice. It has reasonable acid but its purpose is to put colour into otherwise pallid reds. It is a beautiful ornamental vine.

Müllerrebe (German)

See Pinot and Wrotham Pinot.

Spätburgunder (German)

See Pinot Noir.

Trollinger (German)

See Black Hamburg.

AUSTRIAN GRAPES

Austria is a long way to the south of us and its grapes are all far too late to grow here outdoors on any but well favoured sites, but they are available in the U.K.

Jubilaumsrebe C L G (Blau Portuguese X Blaufrankisch)

Intense flavour and bouquet. Ripens in late October in Austria and will require an excellent site or cover.

Zweigeltrebe C L G (Blaufrankisch X St. Laurent)

Very vigorous, a late flowering variety with good thick leaves. At best a blending grape, but favoured for its good yield. Southern Counties only.

DUTCH GRAPE

Roem van Boskoop (S) C M/L

A modern early Sweetwater grape of good dessert quality. Ripens in early October. Likely to be better under glass or on a wall.

AMERICAN VINES

These are all suitable for Cane Pruning and ripen cane adequately.

Brant

A Canadian variety with beautiful golden, red and brown Autumn colours which make it an ideal decorative wall grape. It will set an abundance of black fruit which are good as dessert. Its wine needs a couple of years bottle maturing to make a wine of fair to poor quality.

Concord

A grape with beautiful Autumn foliage, which is really all that it has in its favour, for it is very late and makes quite unpleasantly 'foxy' wine. If it crops then the fruit is of fair eating quality.

Campbell's Early

Happy on well drained soil. Produces large pleasantly flavoured eating grapes. Makes good 'raisins'.

Himrod Seedless (Ontario X Thompson's Seedless)

A well flavoured dessert grape which dries well. A reliable cropper on a warm wall.

Schuyler (Zinfandel X Ontario)

This grape will ripen during November in the U.K., and so needs protection. Recommended for wall culture in the South. This is a very popular grape in the U.S.A. and makes a Zinfandel-like wine.

RUSSIAN V. amurensis HYBRIDS

The three which are listed are said to have Pinot Noir as the female parent. They crop well if late and so East Anglia is likely to be as far north as they will ripen reliably. Elsewhere they should be planted against a wall or under cover. I confess a preference for non-hybrid varieties, but rationalise that 'Amur' is at least part of the same land mass as Europe and that European/amurensis crosses will be without foxy taint. In fact they are very pleasant.

Campbell's Early and Mrs. Pince

Tereschkova

Gagarin Blue

Black Hamburg

Chaouch

Chasselas Rose (Gutedel)

Madeleine Royale

Mrs. Pince's Black Muscat

Gagarin Blue C L G

This is the softest of the three, with pale thin leaves. It produces large handsome open bunches of fine blue-black berries which need no thinning for dessert grape production. If grown for its well coloured wine then it must be destemmed before crushing or the wine will be unpleasantly tannic.It can be prone to stem botrytis in a poor autumn.

Valentina Tereschkova C L G

A grape with a hint of Muscat. It is a hardier grape than Gagarin with smaller red-black grapes of fine eating quality. It is very disease resistant. Like Gagarian it makes a fair wine of good colour.

Kuibishevski C L G

The hardiest of the three, Kuibishevski sets very large bunches of deep red fruits. Its wood ripens well to a rich red. Very disease free.

The National Collection of around ninety varieties of Outdoor Grapes is at Yearlstone Vineyard, Bickleigh, Devon. Yearlstone (8 acres) is one of the few English vineyards which grow red wine varieties.

B. R. Edwards of Pontrilas sells from a collection of over 100 indoor and outdoor varieties.

DESSERT AND WINE GRAPES FOR THE GREENHOUSE OR TUNNEL

Greenhouse grapes are classified as Sweetwater (S), Muscat (M), or Vinous (V).

Sweetwater are the earliest and will ripen in a coldhouse.

Muscat are of finer flavour and will appreciate help with pollination and a minimum 45° Fahrenheit (7-8° Celsius) up to picking time. They will then keep in a cool dark place for some time. Some of the earlier Muscats will ripen satisfactorily in a coldhouse in East Anglia.

Vinous varieties are the latest to ripen and so need a longer period above 45°F. They are mostly 'Victorian' and were developed to ensure a succession of table grapes for rich households well into the following year. If the cost of heating can be afforded then they crop well and are a testimony to the great skills of the richer Victorians' Gardeners.

A method for keeping grapes is described in a later chapter.

There are two 'easy' dessert grapes for the coldhouse.

Black Hamburg S Trollinger (Germany) Schiava (Italy)

Quite the easiest to grow. This is most widely grown in the British Isles. It is problem free, having tough leaves and it sets well with its own pollen. It is early yet it may have bunches ripening late in the year. Bunches will not hang for long once they are ripe because the fruit is thin skinned and it is not recommended for storing. It is grown as a winegrape in Northern Italy and Southern Germany but results in flabby and uninteresting wine here. If you only have room for one variety, then this is the one.

The Chasselas S
...d'Or ...de Fontainbleau, ...Rose ...Rose Royal Gutedal (Germany)

The Chasselas are ancient grapes but quite easy to grow, producing large, variously coloured grapes but with similar taste. The bunches are handsome, tapering and with good shoulders. The ripe grapes have a fine flavour and a beautiful scent. Like Hamburg they set well with their own pollen. The second grape for a collection.

Grapes which are not as easy but which will yield good crops of dessert grapes in the coldhouse include...

Chaouch S

A Turkish golden grape of some antiquity which bears large "'sugar almond' shaped grapes which are very juicy, sweet and highly perfumed.

It is still grown in Anatolia for dessert and to produce 'Chaouch', a dessert wine.

Said to have been a favourite of 19th Century Sultans.

Madeleine Royale S

An ancient grape and so fruiting at the sixth bud but it has been the mother of many good crosses. It is a fair winegrape and good to eat.

Muscat of Alexandria M
and its clone... **Cannon Hall Muscat** (The Guernsey Grape)

This is a classic golden grape, setting perfectly shaped bunches of excellent golden 'Muscat' grapes of beautiful flavour and scent. It is the Lexia raisin of Australia. It sets badly on its own pollen and to the north of the Wash-Galway Line it will appreciate a little extra warmth to get it going and to help it to finish. Well worth the effort. The clone has bigger fruit and a better flavour.

Muscat Hamburg M

A beautiful maroon grape. Well worth growing in every respect. It is a choosy grape which refuses to set a good crop with its own pollen. This most delicious grape will keep in a cool dark place well into the New Year.

Mrs. Pince's Black Muscat M

This grape will set large cylindrical bunches of beautifully flavoured fruit. It is said to need heat to finish but it ripens regularly in Suffolk in a cold vinehouse.

This collection will give a succession of well flavoured grapes, over several months and a fair wine grape of reliable habit. There are dozens of other dessert grapes which are worthy of cultivation in the greenhouse. Some will ripen in coldhouses and some will not.

Read's of Hales Hall, Loddon, have the National Collection of Greenhouse Grapes and offer around fifty varieties for sale.

HOTHOUSE VARIETIES

Alicante V

Will produce in the coldhouse but heat improves its flavour most considerably.

Appley Towers V

A very large ovate black and richly flavoured grape of excellent repute.

Buckland Sweetwater S

An old and totally insipid 'British' wallgrape, included here for its historical interest... suffered badly in the 18th Century powdery mildew epidemic.

Gros Colmar V

A good grafting stock for Muscat Hamburg in encouraging it to set. Otherwise requiring more heat than it is worth.

Lady Downes Seedling V

A richly flavoured grape which needs heat to finish.

Madresfield Court M

A fine tender Muscat flavoured grape, which is elegant in every respect and whose bunches can be very large indeed. It must have heat for perfection.

Mrs. Pearson M

Golden grapes with an excellent taste. A strong grower but late to ripen.

A dessert grape seldom makes good wine for it lacks sufficient acid to balance the alcohol. However, wine grapes can make good eating but they have, usually, smaller fruit and so seem 'pippy'. This fault can be corrected by selecting certain bunches for dessert and thinning the fruit as one would with dessert grapes in the greenhouse. Most purely dessert grapes need cover and the selection of winegrapes is greatly extended by having glasshouses or polytunnels.

RECOMMENDATIONS

It is my opinion that American hybrids are not worth space as winegrapes, but some Russian Asiatic hybrids are useful for wine, dessert and ornament.

There are many varieties which can be successfully grown here and so the gardener may suit his or her palate.

The varieties which I list below suit our palates and some of them have been producing table and winegrapes here for more than 25 years.

Outdoor White

Siegerrebe, Schönburger (both splendid), Précoce de Malingre, Kerner (very late), Müller-Thurgau (not recommended – troublesome).

Outdoor Black

Gagarin Blue, Valentina Tereschkova.

Wall

Perle de Czaba, Tereshkova.

Coldhouse

(Tereschkova, Gagarin, Dornfelder – dessert and frozen for blending with outdoor grapes), Madeleine Royale, Black Hamburg, Muscat Hamburg, Muscat of Alexandria, Gutedel, Mrs. Pince, Chaouche, Siegerrebe (dessert and frozen to blend with outdoor grapes).

ORNAMENTAL VINES

Vitis Amurensis

A vigorous and tough vine from the Amur River in Manchuria (approx. 140E, 54N). The new growth and the large leaves are a bright pink. The leaves turn red and purple in the Autumn. The black fruit is inedible. A parent of some good hybrid fruit varieties.

Vitis betulifolia (China)

A strong growing vine whose 2 to 4" leaves are covered in whitish fluff when young and become richly coloured in the Autumn.

Vitis 'Brant' Clinton (V. Labrusca X riparia) X V. vinifera
var. "Black St. Peter's"

A Canadian bred vine which may grow to be 9 metres high. It bears pleasantly sweet grapes which make a most unworthwhile wine. It is well worth growing for its spectacular Autumn blaze when the large leaves become red and purple with yellow veins.

Vitis californica (Western U.S.A.)

A vine which will be happy to scramble up to 15 metres in a tree. It has soft grey and crimson foliage.

Vitis coignetiae (Japan, Manchuria)

This may be the finest of all ornamental vines. It has broadly ovate leaves may be up to a foot (30cm) across and have a rust coloured fluff beneath. The leaves turn to magnificent maroon and wine in the Autumn and it has 12mm black inedible fruit. It is most striking when scrambling through a tree.

Vitis davidii and **var. cyanocarpa** (China)

A spiny scrambler with 10-25cm shining green heart-shaped leaves which are spiny on the underside. The variety **cyanocarpa** is less spiny. Its black fruits are edible and it turns to a rich crimson in the Autumn.

Vitis pulchra Possibly V. coignetiae X amurensis

A handsome, hardy and vigorous vine. Its young leaves are reddish becoming brilliant scarlet.

Vitis riparia (Eastern North America)

The 'Riverbank grape'. A tall vigorous vine whose male flowers smell of Mignonette. The leaves are glossy green.

Some fruiting vines make good foliage plants, for examples…

Vitis vinifera (Orginally Asia Minor/Caucasus… 'European')

Cultivars – all with edible fruit…

'Apifolia' the 'Parsley Vine' has deeply divided leaves.

'Fragola' the 'Strawberry flavoured Vine'.

'Incana' the 'Dusty Miller' (Wrotham Pinot) with purple leaves covered in floury down.

'Purpurea' the 'Teinturier Grape' with claret leaves which become deep red. Hillier's recommend growing it through a Weeping Pear (Pyrus salicifolia 'Pendula').

Dunkelfelder and Deckrot the German 'colouring' grapes with red juice have fine grey-maroon leaves quite early in the season and the Russian Tereschkova gives a blaze of Autumn tints.

ROOTSTOCKS

When it was discovered that many American varieties were resistant to phylloxera, French and German growers thought that they could 'breed in' the resistance by crossing them with their own varieties.

They were in many instances, correct but unfortunately they bred in 'foxiness' too and the crossings were inferior to their European parent in all respects but phylloxera resistance. However the breeding experiments resulted in a number of good grafting stocks suitable for a variety of soils.

Most of the stocks which are used today are hybrids of two from three native American vines: **V. riperia** 'the riverside vine'; **V. rupestris** 'the rock vine' and **V. Berlandieri** 'the Texas vine'.

V. riperia is a strong grower and encourages early ripening. It has good resistance to phylloxera and tolerates wet conditions.

V. rupestris has similar traits except that it is tolerant to drought and wet.

V. Berlandieri tends to delay and extend the ripening time and will thrive on lean limy soil.

The most commonly available in the U.K. are...

SO4 Berlandieri X riperia which is of medium vigour, is very resistant, tolerant of about 20% lime, encourages early ripening and is suitable for clay-loam.

5BB which has the same parents and attributes, but is a little more tolerant of lime and prefers poor soil.

125AA with again the same parents and very happy in a wide range of soils.

5C is another similar cross but with the most tolerances of limy ground.

3309 is a V. riperia-Berlandiera cross which does well on soils which are liable to dry out easily.

The first four are the most used in Britain.

"That simple dictum...
Light meat – white wine, dark meat – red wine... is simply, not true".
Erwein Count Matuschka

Chapter Two Summary

- ★ About 100 varieties of outdoor grapes will grow south of a line The Wash to Galway.
- ★ The later ripening outdoor varieties will do better on a South wall, in pots or under cold cover.
- ★ Wine grapes are generally good as dessert.
- ★ Dessert grapes generally make poor, flabby wines.
- ★ The best dessert varieties need cover and some need heat.
- ★ The author's recommendations…

Outdoor White Wine/Dessert
Siegerrebe, Schönburger, (Kerner/Kernling on a wall)

Black Wine/Dessert
Tereschkova, Gagarin Blue

Coldhouse White Dessert
Muscat of Alexandria, Madeleine Royale.

Black Dessert
Black Hamburg ('easy'), Mrs. Pince, Muscat Hamburg (Dornfelder or Tereschkova for dessert/wine)

South Wall
Perle de Czaba, Kerner/Kernling, Tereschkova.

Ornamental
Vitis coignetiae (Tereschkova gives dessert, wine and Autumn colour).

CHAPTER THREE

Planting, Training, Pruning and Propagation

Selecting and Preparing the Site

The ideal site is early, warm, free draining, with a sloping Southern aspect and having all the necessary nutrients in ideal proportions and with a perfectly matched pH to ensure optimum take-up.

Fortunately the grape is tolerant of many conditions which are much less than perfect, but one or two varieties are totally intolerant of pH values which are opposed to their needs. For example... Chardonnay and Pinot Noir will not thrive outside an alkaline environment and at the other extreme, Siegerrebe will not tolerate an alkaline soil. No grape enjoys growing more than two hundred metres above sea level (600ft) unless there is some compensatory protection such as a walled garden or plastic or glass cover.

It is most unlikely that a vineyard will be the first consideration when planting a garden and so the grape's general tolerance is a great asset.

Choose the best site which you have available and so long as it is capable of growing good garden produce it should serve the vine well.

Most grapes prefer a site which is...

1. Less than 200m above sea level.
2. Has a pH of 6.3 to 6.8.
3. Is in good heart.
4. Is free from perennial weeds.
5. Is not liable to waterlogging.

[illegible] otherwise [illegible] the other occupants
will be robbed of light and air.

that were more practically instructed Beulah to feed me a grape. From time immemorial the grape harvest has been a season of laughter and happiness.
The large private house gardens all included a large vinery specially constructed for the cultivation of that most important of all dessert fruits the grape: Although my first experience of vine culture was of a Black Hamburgh planted on a sheltered south facing wall in my Uncles garden in Norfolk. it was not until I started working in a large private garden that the business of training and pruning was properly explained. Planting out of containers can be done at any time

The Vine was held by the ancients sacred to Bacchus and old historians all ennaced the jovial god with "the life giving tree". Culpeper, in his herbal STATES that the Vine is "a gallant tree of the Sun very sympathetical with the body of man." and that is the reason grapes, or their derivatives are the greatest among all vegetables. IN 1106 BC THE JEWS "WENT OUT INTO THE FIELDS HARVESTED THEIR VINEYARDS. AND TROD THE GRAPES.

Grape seeds found near the sites of stone age settlements in many parts of Europe suggest that vines were spread far and wide by primitive man at the very dawn of civilisation. More recently poets like Dylan Thomas 'broke the grapes

Though vines will tolerate a wide range of different soils the ideal site would provide free drainage and reasonable fertility on a slight slope facing due south and with shelter from cold winds. A pH of just on the acid side of neutral is also to be desired. Given the right conditions growing grapes successfully WHETHER PLANTING INDOORS OR OUT. Prepare the site well by digging in well decayed organic matter, compost, manure or similar humus forming material plus a well balanced fertiliser, FOR EXAMPLE blood, fish and bone meal. Vines can be grown in a GENERAL PURPOSE greenhouse or conservatory though it will be necessary

Planting Out of Doors

Planting may be done at any time from the Autumn until Spring and so far as container vines are concerned, at any time. The safest time for planting bare rooted plants is the Spring especially if the wood is less than pencil thick, for a hard frost may freeze completely through and kill the young vine. If Winter planting is unavoidable then there is advantage in temporarily covering the graft and/or the lower bud with bracken, which must then be removed in the Spring in order that the scion does not root around the graft scar.

The site should be well prepared beforehand by digging or rotavating in well rotted manure or compost and a well balanced fertiliser. Each gardener will have personal views upon whether to use 'Organic' or 'Artificial' fertilisers but generally a suitable base dressing is 100g per metre square (3oz/sq.yd) of either J. I. Base or National Growmore.

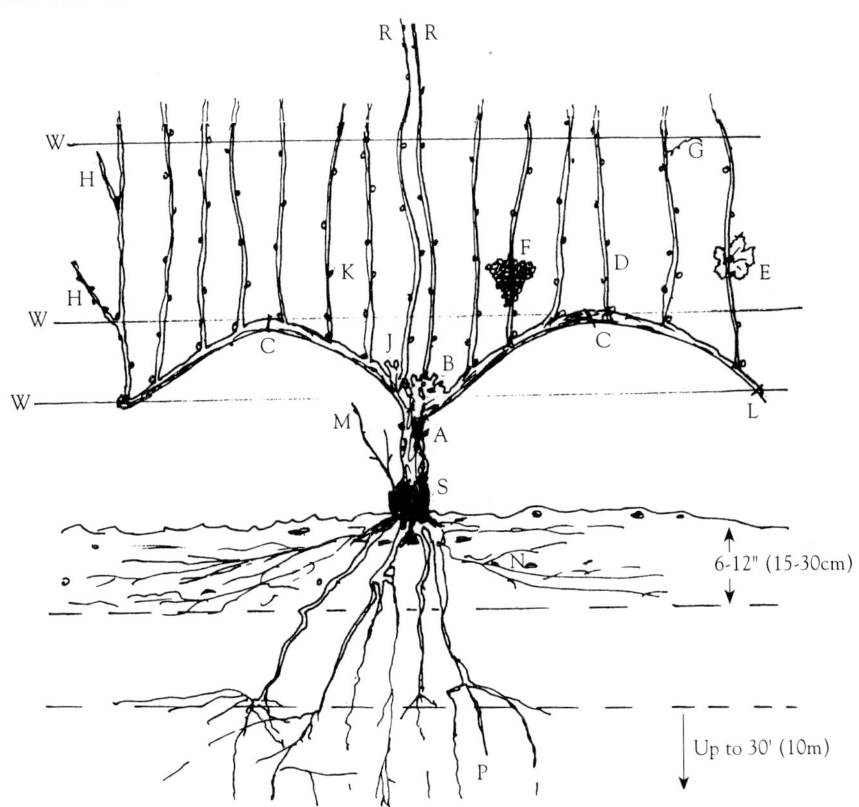

Parts of a Grafted Grape Vine

A. Stem (Scion).
B. Head.
C. Two-year wood.
D. Fruiting lateral.
E. Leaf.
F. Bunch.
G. Tendril.
H. Sub-lateral.
J. Spurs.
K. Bud in leaf axil.
L. Arched wood tied to wire.
M. Water-Shoot.
N. Surface roots.
P. Deep roots.
R. Replacement canes or rods.
S. Rootstock.
W. Training wires.

The vine is a very deep rooted plant – up to 10 metres (30ft) in a gravel environment – and the newly planted vine will soon be into the native sub-soils. There is no long term advantage to be had from 'fussy' planting.

Simply place the vine in a hole of similar diameter to the root spread and deep enough for the vine to be at the same depth as previously and then, fill and firm.

The very worst thing which can be done is to create a water sump by using peat or any other water retainer as the hole filler or planting base.

The vines should be planted about 1.5 metres (4'6") apart with 1.5 metres between the rows.

These distances are not critical, but if the vines are planted too far apart then the vineyard's micro-climate will be impaired.

There is also the disadvantage that strong growing varieties will become rampant if given more space than is sufficient.

A 1.8m (6') cane gives enough support for the first year, but it is better to build the Guyot fence or position the chosen support at planting time and so avoid the possibility of root damage at some later time.

Whichever training system is decided upon, the young vine will be allowed to grow only one cane in the first season and that must be cut down to three or four buds on a 40cm (15-16") leg.

During the first season all side shoots must be pinched out and the growing cane tied in to its support.

It is good practice to use different ties for "Next year's wood" and "This year's wood". For example, orange baler twine for next year's and fillis for tying in fruiting laterals. Whilst it is of no consequence however you tie in for the first couple of years, it is invaluable, from the third year and onwards, to see at a glance "Next year's wood's" ties through an abundance of leaves... and so avoid cutting it out! In fact the first consideration when addressing the vine is... "Which is next year's wood?"... and having found it, satisfy its needs before considering the potential crop.

TRAINING AND PRUNING OF OUTDOOR VINES

There are illustrations of all systems which are described.

Guyot Training

The Single and Double Guyot are considered to be the 'Traditional' training systems and these will be described first of all, but with the Author's reservation that he does not believe them to be, necessarily, of great convenience to the gardener nor advantageous to the garden vine.

They are the most useful to professional growers for they lend themselves to mechanical cultivation and harvesting but I cannot understand why they should be thought of as particularly useful in the garden or in the small vineyard and for the following reasons...

1. Each row requires at least two stout straining posts.
2. Each post requires strong anchoring.
3. Each row requires 5 sets of wires, 3 of them double.
4. It is not conveniently possible to work around the vine.
5. If there are two or more rows, it is not possible to cross the vineyard without going to the row's end.

The Guyot System requires end posts and strained 14 gauge (minimum) training wires singly at... 35 and 42cm (15 and 21") and the three pairs of wires at 90, 120 and 150cm (3, 4 and 5'), the vines are planted each with its supporting cane and then cut back to three buds.

If the rows are much longer than eight metres (25') then extra posts will be required at approximately 5 metre (15') intervals.

The vine is allowed to grow one rod during the **First year** and when this rod has been selected the others must be rubbed out. At the end of the first year and not later than January, the single rod is cut down to three of four buds at about the height of the lowest wire.

A dressing of J. I. Base or Growmore should be given in the Spring of the **Second year** at about 1oz per yard run (30g/m) and the vine allowed to grow two rods.

Any laterals are stopped at one leaf. In this way next year's fruiting buds are fed and the vine does not waste energy in useless growth.

In late August when the rods will have reached 4-5' (1.5m) and should be stopped.

At the end of the second season and before the end of January the poorer rod is cut down to three buds to form a spur from which the fourth year's replacement rods will grow and the better one is cut back to ripe (nut brown) wood and to not more than eight upwardly facing buds.

This cane is arched up to and tied to the second wire and its tip tied to the lower wire. If neither rod has reached 4' then the second year regime should be repeated.

Quite obviously this is as late as you can delay constructing the Guyot fence.

From now on, this will be the time of year when the gardener can shape the vine. It must always be done with a view beyond the coming season. The vine is now

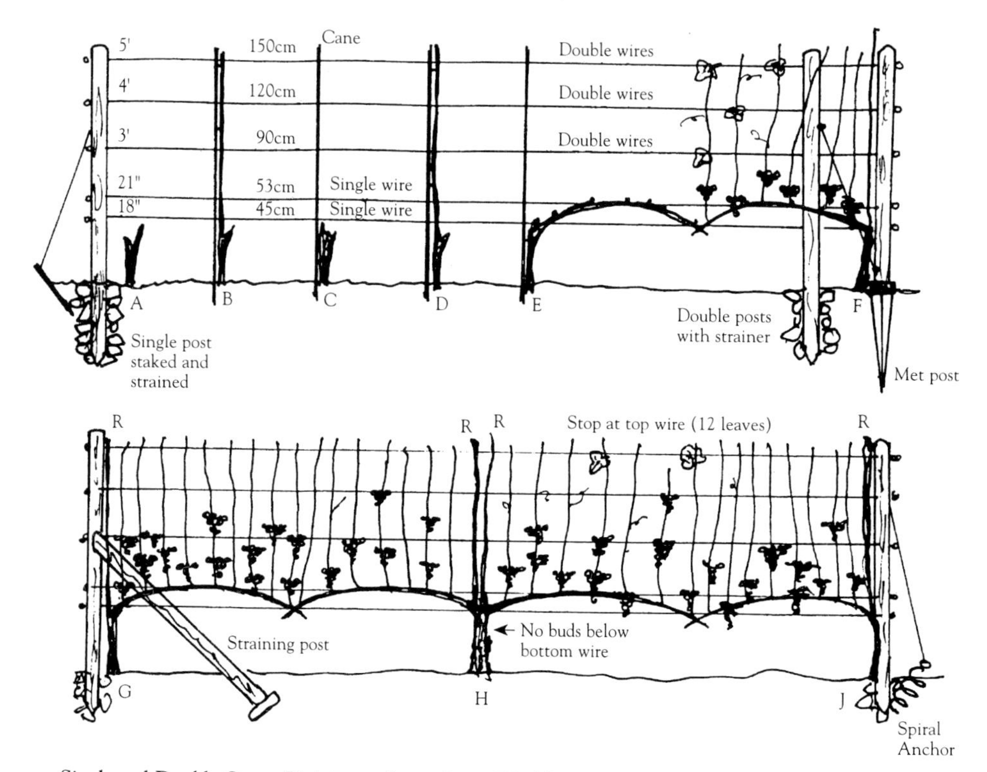

Single and Double Guyot Training and a variety of End Posts
(The majority of leaves have been omitted for clarity.)

A. Cut to two buds after planting.
B. First season's growth.
C. End of first season.
D. The second season's growth is...
E. tied in as 'Single Guyot' to grow...
F. the first light crop in the third season
G, H & J. Mature Single and Double Guyot may have up to two bunches per fruiting rod.
R. Replacement canes (which are best left unfruited.)

Single Guyot Trained, which training ensures that the buds nearest to the head of the stock are not missed by the surging Spring sap.

As the rods grow they should be trained between the pairs of wires.

Some growers carry out occasional fungicide spraying during the first and second years as a guard against infection but the Author considers it preferable to concentrate on keeping the vineyard free of weeds and to supplement the vine's diet with regular sprays of seaweed extract... and so growing a plant with some natural resistance.

Regrettably, although such practice must help, it is seldom sufficient protection when the vine is carrying flowerbuds or fruit. Indeed it is not worth taking the chance that vines will remain fungus free unaided and a regular spraying programme should commence in the third year.

The third year, is the year which rewards your good gardening practice!

It is also the year when overcropping may ruin the vine for many years!

The buds will change in appearance during the latter part of April or early May and will then grow rapidly.

Bud-burst

The rapidly growing fruiting rods will soon show whether they are to carry fruit by the appearance, or not, of embryo bunches.

Firstly, two buds are selected for the Fourth year, either on a spur or growing directly from the head.

Then the objective is to select up to seven conveniently placed fruiting rods which show embryo trusses and then rub out the rest.

There is never any useful purpose for extra and unfruitful growth.

These chosen rods may produce second trusses of flowers and these may be allowed to develop, but it is recommended that they are not... in the vine's future interests seven bunches is a crop... and most certainly a vine, however strong it may look should not be allowed to carry more than seven bunches at the first harvest, for even with such a limited crop the juice and flavour will be thinner than in future years.

On no account should a third truss be left on a rod and it must be removed whilst it is tiny for much energy is expended in flowering.

It should be noted that the 'Fruiting and Replacement Rods' which rise from the Two Year Wood and the Head are alternatively known as 'Laterals' and are quite clearly seen as "lateral" when the vine is greenhouse grown for they will then grow horizontally from the spurs on the vertical Permanent Rods.

The spray programme should begin when the leaves are about 1.5" (4cm) in diameter. (See later chapter)

Replacement Rods

Many growers allow three to grow from the spur of the head. Two to form the Double Guyot training for the fourth year and one to be cut back (as at the end of the second year) to provide canes for the year after.

Upon the premise that only useful growth should be allowed the Author believes that two rods suffice, for there are always buds enough to choose from at or near the head.

From the end of the fourth year and onwards the next year's Double Guyot is arranged by cutting out the spent wood back to the two rods most conveniently nearest to the head.

Having made that choice it is imperative that next year's canes are not pruned inadvertently when they sprout from the head. This was my reason for suggesting the use of differently coloured ties for differing purposes and so next year's rods should be tied loosely to the support with some brightly coloured twine as they grow.

Laterals, Care of Replacement Rods and Stopping

Any laterals growing from the chosen replacement rods should be pinched out at one leaf and this leaf will guard and feed next year's fruitbud.

Alternatively, these rods may be cropped together with six others on each of the two year fruiting arches. This is not a recommendation for it has the disadvantage that the rods tend to develop spurs and make too many buds.

All the sub-laterals on the fruiting rods should be rubbed out as soon as they are seen and throughout the season the gardener must 'stay on top' by rubbing out laterals... or else the vine will become a thicket of laterals and sub-laterals, useless foliage and little to show for a crop. One regular visit a week will suffice, but you will find yourself in the vineyard more often than that!

The fruiting rods are stopped as soon as they pass the top pair of wires, when the rod has 12-14 leaves. This allows the vine to use its energies to ripen the fruit. The leaves are photosythesising sugars and these sugars should not be wasted on unproductive growth.

As harvest-time approaches the lower leaves will take on Autumn tints and their removal will allow more ripening sunshine onto the fruit. (See later chapter)

The Fourth and Subsequent Years

At the end of the third year the vine has a Single Guyot arch and the spent fruited rods.

Firstly be sure of at least two buds at or near the leg or cut down the weakest of the two or three replacement rods to three or four buds.

Next cut out the whole of the recently fruited wood to as near to the head as is possible.

Thirdly, and with reference to neighbouring vines, arch the two replacement canes to each side of the vine and tie them to the lower and second wires as was done for Single Guyot. Again cut back to about eight upwardly pointing buds.

The vine is now Double Guyot Trained.

You will have discovered that the vines stationed at the endposts will always be Single Guyot trained and these may carry up to ten rods and a replacement rod.

Always look beyond the coming crop... if a vine has grown poorly, then prune it severely. If Double Guyot then revert to Single Guyot for the next season, or allow fewer rods per arch.

Guyot Training Ancient Varieties

You will have read that some ancient varieties will fail unless spur pruned, because their first fruiting lateral rises from the sixth bud.

If you are growing, for example, Précoce de Malingre then from the Single Guyot year onwards the arches are permanent and the rods are pruned to eight buds after cropping.

At budburst leave the first bud on the spur but rub out all other buds up to the first growth **showing an embryo bunch** on the sixth shoot (or 7th or 8th which are 'insurance' buds). The first bud is stopped at eight leaves for next year's spur. You will cut out the old spur after cropping.

There will come a time when the arches begin to look gnarled with old scars and spur stumps. This can be dealt with by growing a new rod whose laterals are stopped at eight leaves... this will allow the replacement of one arch after harvest... and the other arch can be attended to in the same way for the following year.

This system may be modified and used for varieties which tend to fruit at the third bud… for example Reichensteiner. In these cases the spurs are pruned to five buds.

You may wonder why you should not use this permanent Guyot for all other varieties… it may be done by spur pruning the fruited rods to two or more buds (in effect ensuring a short length of 'two year' wood) and this **is** the practice in Bergerac where the vines are grown as single cordons with up to a dozen fruiting canes, each with a spur to make two year wood. It amounts to no more than a matter of usual practice.

Mosel 'Hertzliche' (Heart-like) System

This is a modified Guyot System and one which allows access all around the vine.

It was developed for use on the very steep vineyards of the Mosel but it is most useful in the garden.

It requires that each vine has a stout stake at least 5'6" (1.8m) above ground. I use 1" square stakes which are tied at 6'6", to a line which runs from end to end of the fruit cage.

In the Mosel the stakes are freestanding and are about 3" (8cm) in diameter above the point.

The procedure is as for Guyot training, except that the vine is on a 2' (60cm) leg and the Two Year wood is arched right around and secured to the leg so that in the fourth year and onward the trained vine has the appearance of a heart. This deep arching of the wood guarantees that the buds are fed in turn by the Spring surge of sap.

The two Replacement canes are grown to at least 5' (150cm) and tied to the stake with coloured string.

As the fruiting rods grow they are loosely anchored to the stake with fillis.

I find it convenient to loop around all the rods in one.

As the season progresses the system becomes increasingly self supporting.

It is imperative that the grower attends to removing sub-laterals, for the rods are quite close. Any laxity will result in a jungle and so each vine must be visited once a week. Six or seven bunches is plenty for the vine in its third year (…when it is 'Half-hearted') and up to fourteen in its fourth year and subsequent years and a rod should never carry more than two bunches or the vine will suffer.

Mosel half-trained

Moseller Hertzliche (heart-shaped) trained

Das Trierrad (Trier Wheel) trained (3rd year)

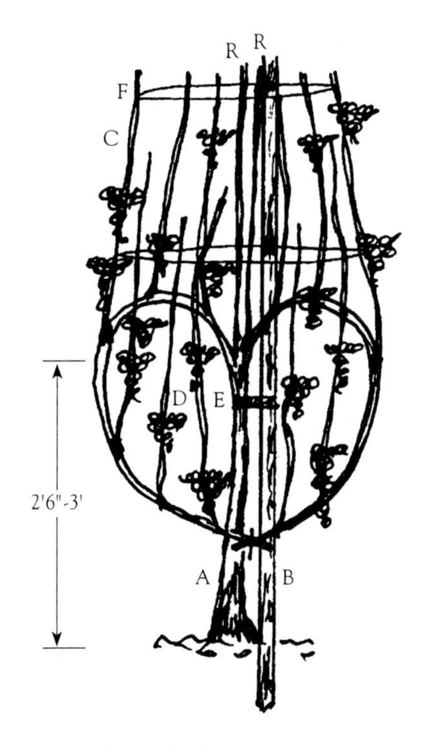

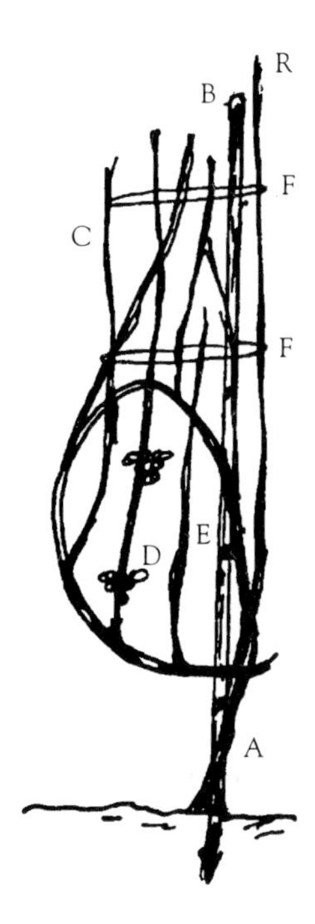

Mosel Hertzliche Training and Half-Heart Trained
A modified Guyot System but on an eventually higher head.

A. Stem, eventually 2'6"–3' high.
B. Post.
C. Fruiting lateral.
D. Eventually up to two bunches per lateral.
E. Stem tie.
F. Loose loops around laterals.
R. One or two replacement canes.

There are differing opinions on exactly how many leaves should be left beyond the last bunch regardless of the training system… 'past the top wires' (Guyot) is about 12 to 14 leaves per rod or 6 to 8 beyond the second bunch in all cases. Nick Poulter in 'Growing Vines' considers that… "under no circumstances should outdoor vines be stopped a few leaves beyond the bunch… or the bunch may be thrown… and also the leaves are 'factories'."

Some older and more recent references recommend pruning to two leaves. This recommendation must work for some but I would be concerned that the bunch might abort, or if not, that there was not enough 'factory' for unstressed growing and ripening.

Vines must grow sufficient foliage to sustain themselves **and** to ripen their fruit and to these ends, 6 to 8 leaves beyond the last bunch strikes a good balance out of doors.

The fruited wood is pruned back to the ripe replacement rod(s) each year and the new 'heart' is made.

Das Trierrad (The Trier Wheel) System

This system uses gravity to take the rods and fruit down from ten or a dozen buds atop a 5' (1.5m) standard.

The head is built to a convenient number of spurs over the years and the fruiting laterals are arranged evenly over a 'wheel' (rather like a lorry steering wheel) of about 18" (45cm) diameter which is fixed by its boss to a stout stake.

The Trier Wheel is of moulded plastic and designed to fit over the top of a standard German vine stake.

I make 'Trier wheels' from 55" (1.3m) lengths of 12mm polypropylene hot water pipe joined as 'wheels' by lengths of 12mm copper pipe. Two diametrically arranged lengths of batten form four spokes which are galvanised nailed to the wheel… one within the diameter and the other at right angles and beneath the wheel. The wheel is fixed to a stout stake by a 4" galvanised nail driven through the drilled spokes.

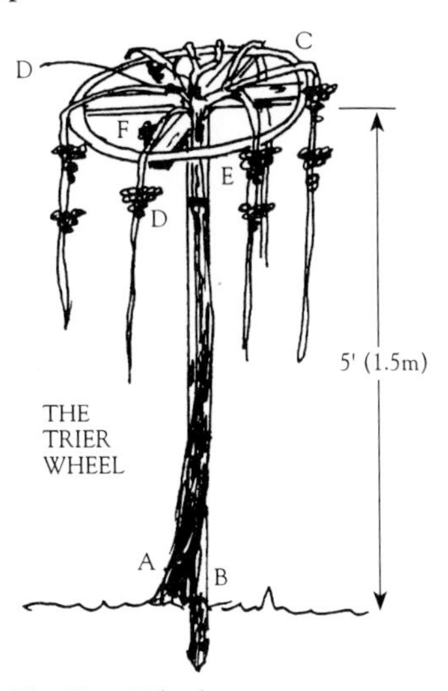

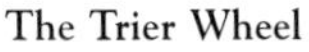

The Trier Wheel

A. Stem.
B. Stake.
C. Alkathene Wheel.
D. Spur-pruned Head.
E. Two bunches per fruiting lateral.
F. Batterning spokes.

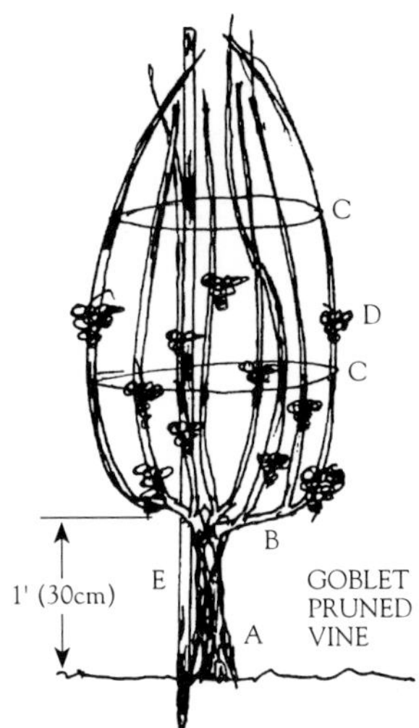

Goblet Pruned Vine

A. Stem.
B. Spur-pruned Head.
C. Loose Loops.
D. Up to two bunches per fruiting lateral.
E. Post (may not be required).

This system is most useful at the South end of a row of Mosel trained vines... for it casts little shadow on a neighbouring Heart-trained vine... or between the last vines of two rows in a fruit cage.

In the early years the vine may be Mosel Heart-trained and changed to Trier Wheel by using one length of two year wood as the standard.

Goblet Training

This is also a spur system, but in this case the rods are trained upwards from a short leg and after the first couple of years, when it will need a cane, the vine becomes a free standing bush.

The first year training is the same as for all other systems.

In the second year the vine is allowed to reach about 1' (30cm) and stopped, when all buds but the top two should be rubbed out.

These will grow and they too are stopped at two leaves.

The vine is then left alone to grow as it will, but it should be gathered into the cane to protect against strong winds. At pruning time all the ripe growth is cut back to six buds, to accommodate ancient varieties such as Perle de Czaba, Gamay and Précoce de Malingre, or to three buds for most others. Eventually the 'Goblet' will have a maximum of twelve fruiting spurs which will be allowed to produce twelve fruiting rods and up to twenty four bunches.

As the season progresses it will be necessary to circle the rods with fillis and to stop them at 12-14 leaves.

The Geneva Double Curtain

This is a High Trained System developed in New York State and it is intended to be grassed down.

It is favoured by a few U.K. commercial growers situated, mainly, in the South.

It requires very large and stable end-posts and very heavy gauge training wires and it is essential that your chosen variety has the potential to grow well enough to suit this high training.

Perhaps its most likely use to the gardener is in a modified form as, or akin to, a Pergola.

In its original form the end-posts are 10" (25cm) square and 6'6" (2m) high with another 3' (1m) below ground. These are made into "Tees" by stout and triangulated 4'6" (1.5m) cross-pieces. The training wires are strained at about 4' (1.3m) apart. One row of vines is planted between the end-posts 9' (3m) apart.

The vine must firstly be grown to the height of the system when two rods are taken across to the wires... one to each... and then trained along the wires in opposite directions until they reach the next vine.

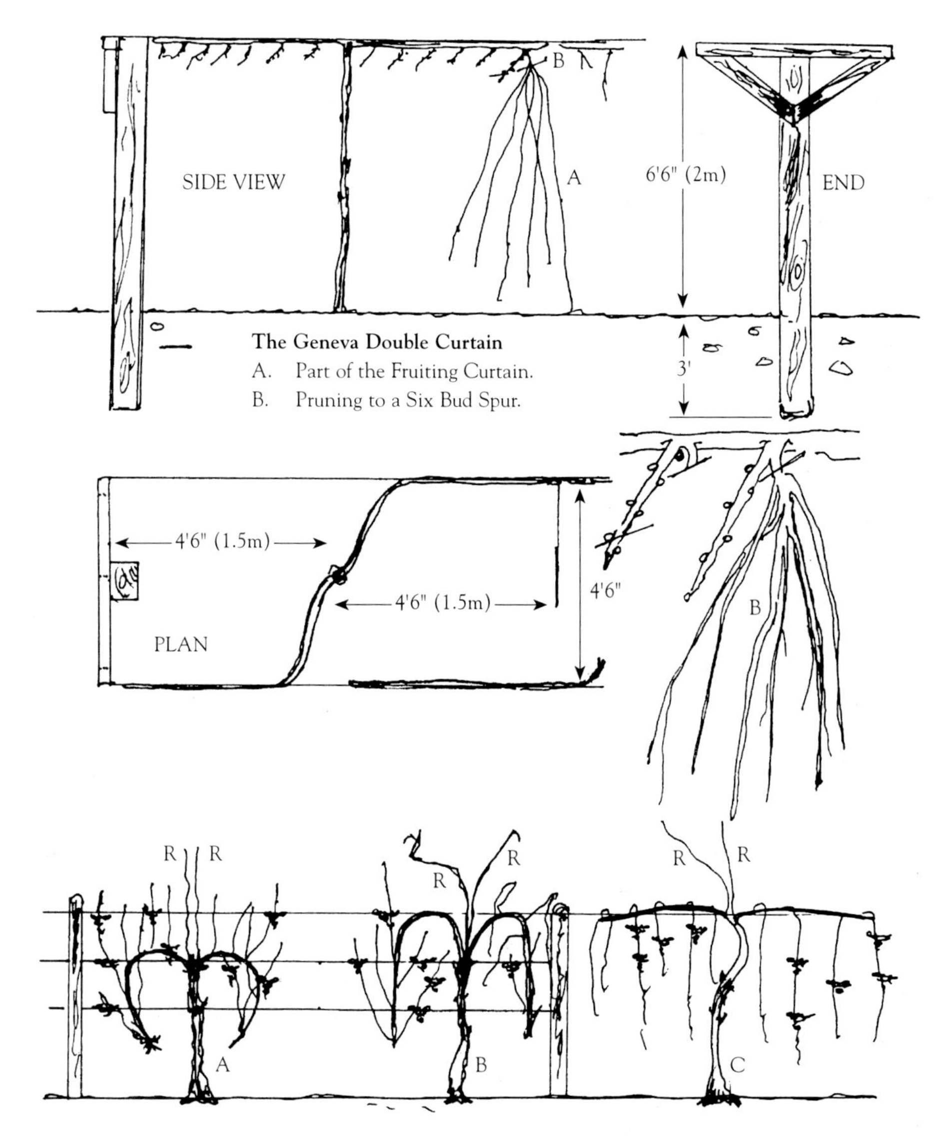

The Geneva Double Curtain

A. Part of the Fruiting Curtain.

B. Pruning to a Six Bud Spur.

A. Alsatian Training.

B. Umbrella Trained.

C. High Double Guyot.

R. Replacement Canes.

Eventually, each rod will be furnished with spurs of lengths suiting the variety and the system lends the appearance of having a 'Double Curtain' of fruiting wood. The leg and rods are permanent, with end of season pruning back to spurs.

It is the Author's opinion that Geneva is a system for specialists with special provenances and probably at its best in the hot humid conditions of the North Eastern United States Summers... for which it was developed... but it may have potential in some modified forms for an enthusiast.

There is a multitude of training systems, but those which have been described will serve most gardeners well or be modified to suit the practised gardener's requirements.

A few Regional systems which are of academic interest to the British grower include...

The Chablis System (much used in the Champagne Region) where the vines are planted 1m x 1m, and are allowed to grow four right-leaning stems from an almost ground level crown. When an outer stem reaches the next vine it is removed at the end of the season and a replacement cane, which has been growing from the left side of the crown, makes up the number. Pinot Noir is spur pruned to four buds and Chardonnay to five buds. Other systems which are used in France include...

Alsatian Training which is an exaggerated Double Guyot System although not as pronounced as Mosel Heart Training. It is on a 1.2m (4') leg and the fruiting wood is tied in upwards on a high trellis... **Umbrella Training** which is, in effect, a severe variant of the Alsatian System... and...

High Double Guyot which is trained from a 5' (150cm) leg onto a single wire from which the fruiting wood hangs as a curtain.

The Loire Low Trellis System is essentially Single Guyot with only 5 fruiting canes and trained on only 2 wires. The vines are planted one metre square and are kept to less than a metre high.

The Valée de la Marne System is a curious 'Double/Single Guyot' System when a single replacement cane is fruited together with the outer fruiting lateral on the otherwise disbudded previous year's arch and on a three wire trellis.

The Bergerac System. A cordon of up to 3 metres is spur pruned to two buds and trained on three wires.

All systems are variations of cane and/or spur-pruning. The principle common to all systems is that this year's fruiting shoots will spring from last year's wood.

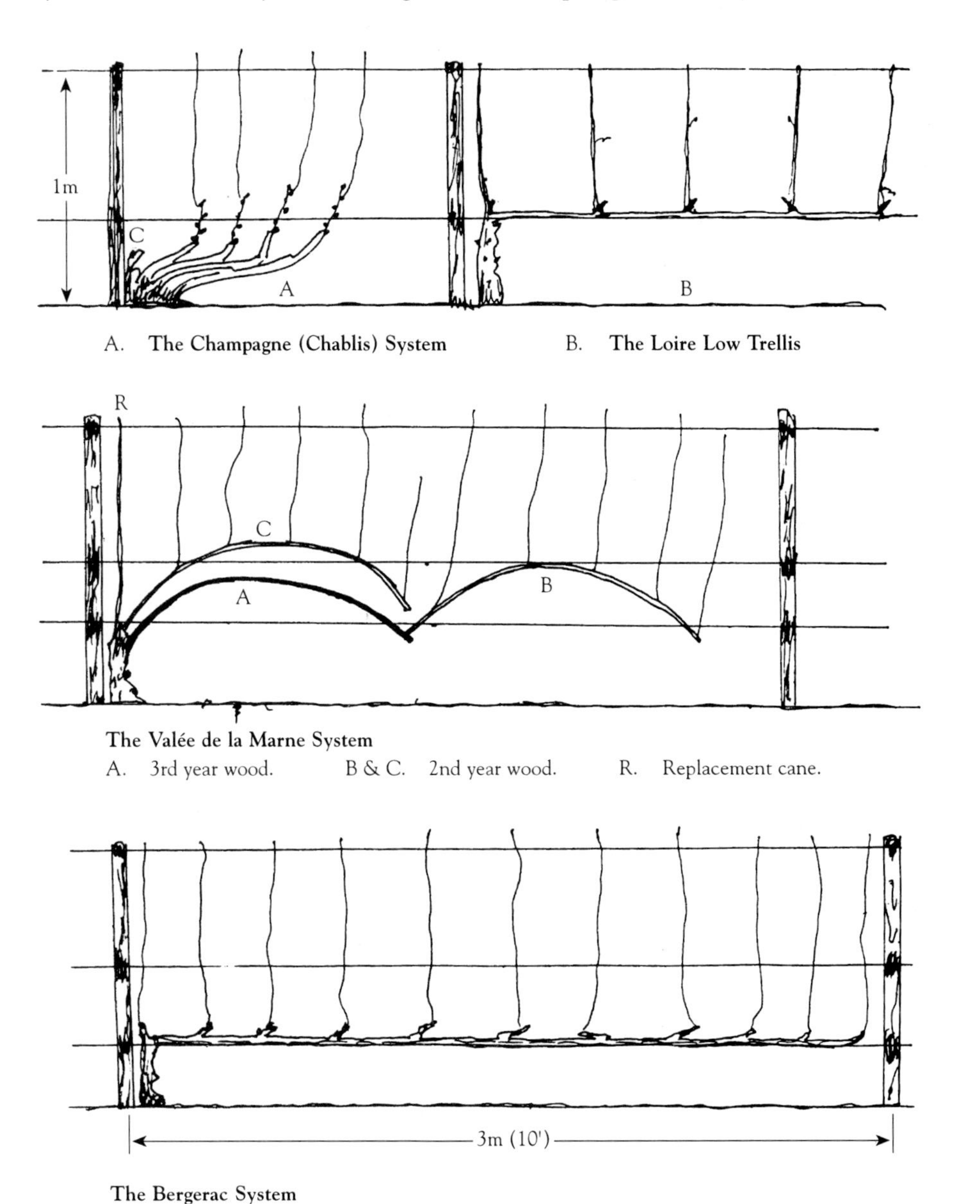

A. **The Champagne (Chablis) System** B. **The Loire Low Trellis**

The Valée de la Marne System
A. 3rd year wood. B & C. 2nd year wood. R. Replacement cane.

The Bergerac System

Guyot, Mosel, Trier and Goblet are the systems most likely to be chosen by the gardener, and of these Mosel and Trier are favoured by the Author together with Goblet training for its minimal cost and usefulness for grapes with special needs.

WALL GRAPES

The Vine is so accommodating that it may be made to grow and fruit as the grower desires but, generally speaking, there are two basic systems for wall training spur pruned cordons. The choice is decided upon by whether the wall has windows and/or, doors or not.

The illustration shows a variety of trainings which have been recorded.

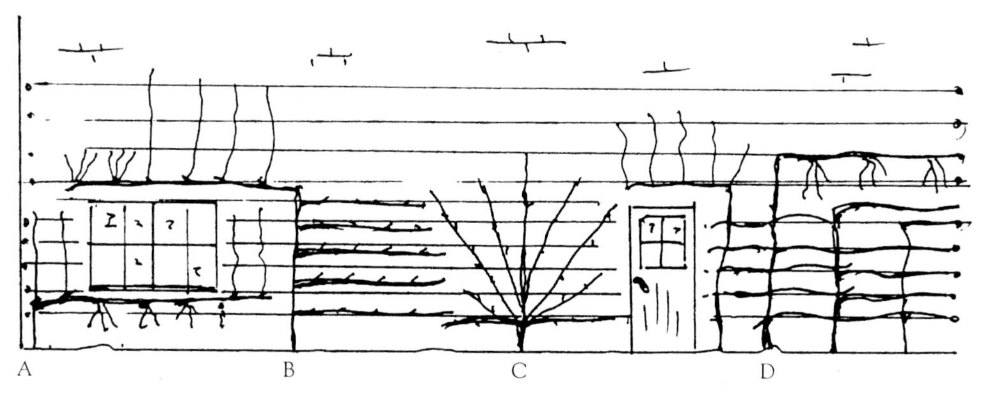

Spur Pruned Wall Grapes

A. Low Cordon.
B. High Cordon and Espalier.
C. Fan Trained.
D. High and Woven Cordons.

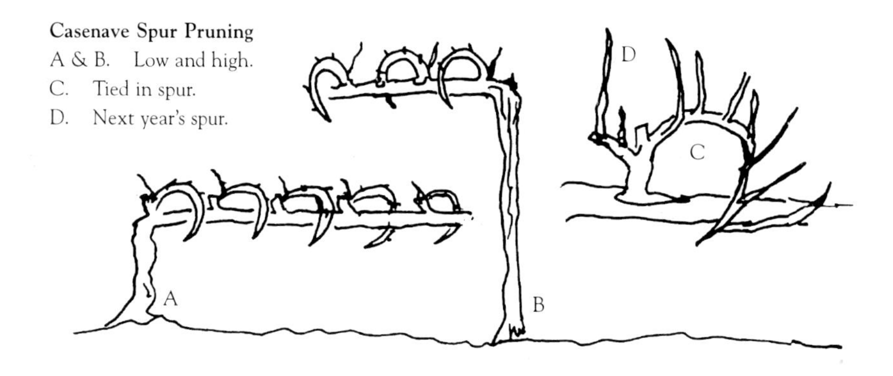

Casenave Spur Pruning

A & B. Low and high.
C. Tied in spur.
D. Next year's spur.

High Trained

If the first year's growth is strong then the cane need not be pruned by two thirds but only to ripe wood. This may or may not be to the height at which the permanent branches are required and if not then a further year will pass before the next training.

When you are able to begin training-in the one or two branches – which will be decided by where along the wall you planted – you will, again, have two choices... Either these permanent laterals will carry fruiting wood upwards, which you will be tying onto wires, one foot apart and 4" (10cm) away from the walls, or the fruiting growth is allowed to hang downwards, but in either case pruning will be done according to the variety's spur pruning needs.

Low Trained

This method is similar to High training and in both systems one foot spaced wires are held by vine eyes which are Rawlplugged into the masonry. It differs only, and obviously, in that if very low on the wall its fruiting wood must be grown and tied-in upwards.

Pruning, cropping and general care are the same as in other modes.

I find it convenient to use Low Training on three wires with a spur pruned permanent cordon... training is a matter of convenience whilst attending to a variety's requirements.

The 'Casenave' cordon is a tidy long spur system which is useful in dealing with ancient varieties whether high trained or low. The spur is tied down to the cordon. The next year the spur is trained in the opposite direction. One or more fruiting laterals can be allowed according to the vigour of the vine.

Other Ways

These include training regimes which are shared with other fruits, for examples... Espalier and Fan training... indeed training is very much a matter of fancy.

If space is very limited, or just from choice... then... **High and Low Double or Single Guyot** are most satisfactory.

Planting and Training Under Cover

The Classic Victorian Vinery was a south facing lean-to greenhouse built against a high brick wall. The south side was a low wall surmounted by side glazed ventilators and then forty-five degree glazing to the high wall with ventilators along the top.

The house was heated by hot water pipes under cast-iron gratings.

The vines were planted either outside and led through the wall or inside and near to the low wall.

The vine was grown horizontally at about wall height and up to a dozen permanent rods grew parallel to and about a foot (30cm) away from the glass.

The rods were anchored to wires, which ran from end to end of the house and spaced at about one foot.

The high house required the use of a step-ladder to attend to the higher growth and crop.

These vines grew massive crops and made great demands for feeding.

A fine example is the ancient Black Hamburg (Trollinger) at Hampton Court which has a large area of lawn outside to attend to the needs of its annual 900 pound (400kg) crop. The rods were cut back to spurs in late December. The heating went on to start the vines in the early Spring and to finish them into late Autumn. The heating was off during the Summer for maximum ventilation and in the Winter when the ventilators were wide open to encourage the vine to rest.

Planting Under Cover

Whether in a lean-to, a ridged glasshouse or a polytunnel, whether heated or not, the presently favoured approach to training vines under cover is to aim to grow two or three permanent rods on a short leg.

The soil must be well prepared and it is best to have been worked over some time

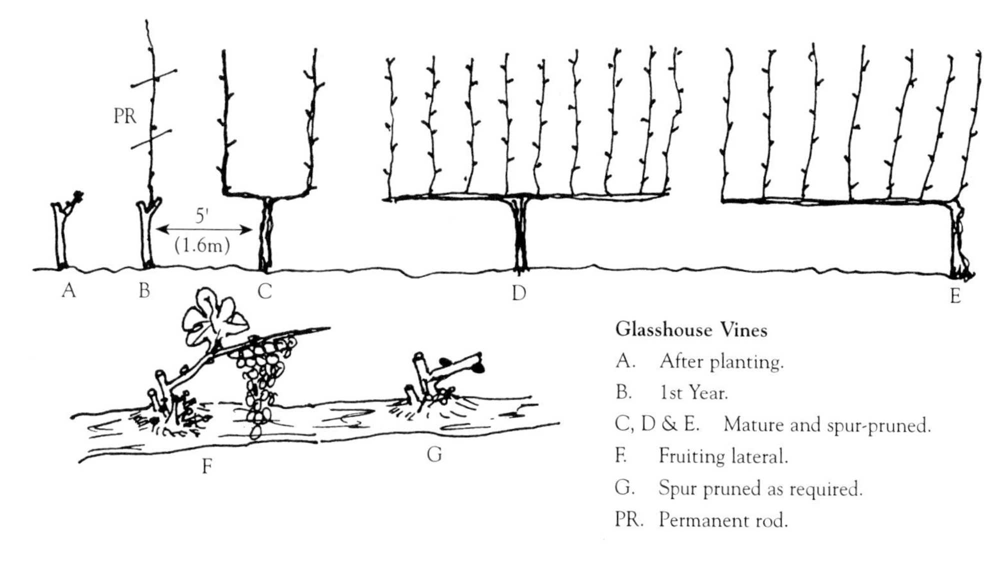

Glasshouse Vines

A. After planting.

B. 1st Year.

C, D & E. Mature and spur-pruned.

F. Fruiting lateral.

G. Spur pruned as required.

PR. Permanent rod.

before planting… like so many other crops, vines do not enjoy being dumped into newly manured land.

Dig in manure, compost and added drainage (broken brick, pea shingle, very sharp sand) if the land tends to pan, a little Borax and Kieserite (Commercial Epsom Salts). A couple of applications over a couple of months will ensure a good mix. This needs to be dug or cultivated in as deeply as possible. In days gone by the whole vinery border was dug out and layered in order with a bottom layer of rubble, then inverted turves, a layer of horse manure and the top layer, of now rather higher greenhouse soil, was replaced.

Very often an aborted lamb or piglet was buried deeply below the vine's station.

In whichever fashion you choose to prepare for planting there is no point in preparing other than the vine border.

There is convenience in setting up the wires before planting. The Victorians' lengthwise system of wires one foot (30cm) apart and one foot from the glass is the best.

Vines in a polytunnel must be stopped two feet (60cm) from the top and that area kept absolutely free from foliage. The one great disadvantage which the polytunnel has, is that it is very difficult to keep ventilated… do have fine net sides and ends. The Victorian growers sank land drains along the path, end on, opposite each vine for Spring feeding with blood.

The First Year

The one or two year-old vines are planted at 4' to 6' (1.3-2m) stations to fit the border and pruned to two or three buds. (Remember that the vines at the ends of the row may be trained around the corners). Some favour root pruning to 4" (10cm) but I do not.

The strongest shoot is allowed to grow unchecked but any laterals should be stopped at one leaf. Two thirds of the growth is cut off not later than late December. In the very unlikely event that little growth has been made then the vine should be pruned hard to a couple of buds and the first year regime repeated.

The Second Year

This is the year when shaping is the prime concern. It is possible to grow two rods and to train them as an elongated "U" with the legs about 2' (60cm) apart, but it is recommended that only one rod is grown, with a spur left for the second rod to be grown in the third year.

It is easier for the vine to ripen one rod and then carry a small crop in year three than to be overworked in its youth. If you intend to have a three rod frame, which can take on a jungle look, then the elongated "W" should also have its rods 2' apart grown one each year. (In the Author's opinion it makes for easier maintenance and allows more variety to have the vines at 4'6" spacing each with two permanent rods.)

All laterals should be stopped at one leaf. It is to be noted that every leaf on a vine guards and feeds the bud in its axil. In this case the bud is, potentially, next year's fruiting wood and so it must be preserved.

The new growth is cut back by two-thirds or to really ripe wood not later than late December and after spraying with tar oil (later chapter) the rods are cut from the wires and the tips laid to the path.

This practice imitates Guyot training and ensures that the buds receive the Spring rush of sap consecutively.

The ventilators should remain open all Winter.

The Third Year

This is the year of the first crop. The vine may well produce fruiting laterals from every bud but you must not be tempted to think that the vine is capable of sustaining more than three bunches on the single rod **and** make the second rod without overtaxing itself.

Even when the vine is older it should not be allowed to carry more than one bunch of fruit per foot (30cm) of permanent rod. This approximates to twelve bunches per vine... perhaps 20lb (10kg) per year by the sixth year from a two rod vine.

Do not be in a great hurry to prune... wait until you can see the embryo bunches.

When you address the vine have in your mind's eye the future... Where would you like the bunches?... How will what you do affect the future shape and cropping?

When you have assessed the vine rub out the blind laterals and then choose two or three of the best looking bunches of buds and pinch out all the others to two leaves.

A small garden vineyard of 48 vines

Outdoor vines pruned for the winter

Coldhouse vines tied in during the spring

Pinch out any further bunches which appear on the chosen laterals and stop the lateral at four to six leaves beyond the bunch. Rub out any sub-lateral buds as they appear and do not allow any further growth on the non-fruiting laterals.

The vine needs supplies of Zinc and Manganese at this time and Boron soon after. (Later chapter)

From budburst and throughout flowering the vinehouse should be kept buoyant by spraying the whole vine and dampening the walls, the borders and the paths daily at about noon.

When the fruit has set the vine should no longer be sprayed, but damping down should continue until ripening begins (Veraison), when watering must cease.

It is most important that water is withheld when the grapes are ripening for, when once the skin is set excess water will burst it and open the fruit to mould or fly and wasp attack.

Subsequent Years

The rods are permanent and so two year wood is provided by spur pruning, usually, to two buds. Some varieties will need more.

From the fourth year and onwards the vine may carry larger and larger crops until, in its sixth year it may be treated as mature and then allowed to carry one bunch per foot of permanent rod.

Rods do age and it is beneficial to grow replacement rods from time to time, the old rod being pruned out in December. There is no need to rush into replacement and two old rods may be replaced over three or four years.

The selection of fruiting laterals and the stopping of blind laterals is done in the same way as previously except that it becomes necessary to train the fruiting laterals alternatively so that bunches are not directly over others.

As more of the wires are covered you must consider adjacent vines as you plan the fruiting structure. This sometimes involves training laterals to the side opposite to that from which the lateral grew. So long as you do this over a number of days it is easily accomplished. If you rush, the whole lateral will come away from the rod and the bunch is lost!

Other Training Systems Under Cover

Permanent rod systems may be arranged to fit both the house and the Grower's wishes. They are all variations of the one which has been described, differing only in the length of the laterally trained canes and the number of permanent rods which they carry.

The grower can, eventually, fill an entire vinery with one vine, shaping it at will.

The pruning will be the same as has already been described.

Setting the Fruit

The traditional way to assist the pollination of dessert grapes was to stroke each floret with a rabbit's or hare's foot making sure that different varieties' pollen was moved onto differing varieties' flowers. An easier and usefully less efficient pollination is achieved by drawing the bunches through your hand each day, making sure that pollen is transferred from one variety to another.

In this way not all flowers are fertilised and so later thinning is easier.

It is essential that the house and the vines are dampened down after this procedure.

Thinning the Bunches

The grapes will have grown to currant size after about four weeks and this is the time to carry out the first thinning. If they were 'hand-fertilised' then many of the inner berries will not have received enough pollen and will have aborted, but some will still remain to be cut out.

This can be the most difficult operation in the growing of dessert grapes. It is necessary to envisage the finished bunch and in order to do so, to have seen a bunch typical of the variety. If this has not been possible then the first attempt may not result in the perfect bunch!

The object is to finish with a well shaped bunch having wide straight shoulders and evenly large berries. This involves removing most of the inner berries (and is done with grape thinning scissors, or long thin hairdresser's scissors). On no account touch the bunches with your hands or you will ruin the bloom… the stemlets may be held away from the scissors with a small forked stick.

The first thinning begins by opening out the centre of the bunch by removing most of the inner berries whilst taking care not to remove berries from inside the shoulders. Then the outer berries and the underside of the shoulders are thinned to leave about 1/4" (6 or 7mm) between each of them.

A second thinning will almost certainly be necessary seven to ten days later.

Each berry will occupy about 1" (25mm) long cylinder of space 1/2" (12mm) in diameter and if this can be kept in mind it will assist in arranging for a well shaped bunch.

There is never any need to thin grapes which are to be used for wine making and one or two varieties... for example Gagarin Blue, make such lax bunches that they seldom require any thinning.

The Victorian Grape Table Decoration

The richer Victorians were very fond of grapes and expected their gardeners to provide grapes for the table for most of the year.

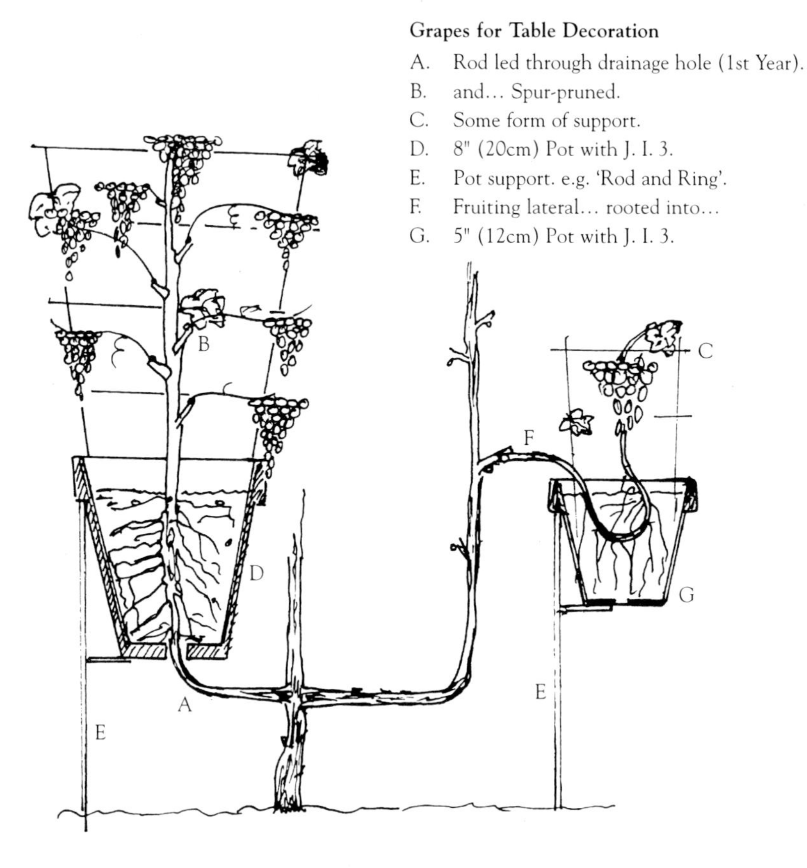

Grapes for Table Decoration

A. Rod led through drainage hole (1st Year).
B. and... Spur-pruned.
C. Some form of support.
D. 8" (20cm) Pot with J. I. 3.
E. Pot support. e.g. 'Rod and Ring'.
F. Fruiting lateral... rooted into...
G. 5" (12cm) Pot with J. I. 3.

Grapes were forced into growth in one house in early January, with bottom heat and provided August supplies, whilst a second larger grapehouse came into growth naturally with the follow-on crops of Sweetwater, Muscat and Vinous varieties. These kept the table supplied straight from the vine until January.

Some bunches of almost ready Vinous grapes were cut with a 'Tee' of fruiting spur and with a longer arm cut nearer to the rod. This arm went into the Grape-bottle which had previously been filled with water and a little charcoal and the bottle placed into a 45 degree rack on the wall of the dark cool grape-cellar. In this manner the table was kept supplied until mid-May. You may 'find' grape bottles but, if not, then milk bottles are as useful.

It was not enough for many households to merely have table grapes... they had to be a part of the ambience along with the Aspidistras and Crotons and the dinner guests were able to pick grapes from a miniature grapevine which appeared to grow in the centre of the table. There were two techniques used by those most ingenious Victorian viticulturalists.

The Rooted Rod

This system takes two years and makes a spectacular pot-plant. A ringed stake supports an eight inch (20cm) pot and a rod is grown through the drain hole. As the rod clears the pot rim the pot is gradually filled with a rich potting compost and the rod grown on and kept pruned as first year wood but trained on a frame.

The next year bunches of grapes arise from the arranged spurs and the pot is kept watered and fed... for the 'one bunch to the foot of rod' is not observed.

At veraison the fruiting rod is severed at the base of the pot, over a number of days, to avoid 'bleeding'.

The Victorians would have insinuated growing Ferns and Moss around the vine before the pot went into, either a decorative outer pot or a fitting hole in the table!

The Rooted Lateral

This is easier.

The system is similar but requires only a five inch (12.5cm) pot and a fruiting lateral. The lateral is looped into the top of the pot so that as much of the early growth as possible may be rooted. The lateral is then trained as though it was a rod and tied in to vertical support. Grow just one bunch.

The lateral may now be left on the vine until the bunch is ripe before cutting away... The vine 'expects' a fruiting lateral to be cut and will not bleed.

The housekeeper would have expected several of these at a time to grace the table.

The Curate's Vinery

This is the Vinery for the small garden and it is both a simple and inexpensive Do-It-Yourself structure.

I am indebted to Read's Nursery for permission to reproduce the design from their catalogue.

The Curate's or Ground Vinery

A. Spur pruned Cordon planted outside.

B. Bunches supported off the flagstones.

C. South-side glazing usually fully open.

The glazing should be of glass or rigid plastic. The Vinery is 7'6" x 3' x 1'8" (2.25m x 1m x 0.5m), triangular in section and with a central ridge. The ends are boarded with the top 4" (10cm) left open for permanent ventilation.

The preferred alignment is as near to North East – South West as is possible. The southerly facing side is hinged to open fully.

The site is made from two rows of 18" square paving slabs set on sharp sand. Two rows of building bricks along the edges, laid half a brick apart, form the ventilated walls, upon which sits the vinery.

The vine is planted in a well prepared bed outside the southerly end and the single permanent rod is led under the end and trained along the centre of the vinery.

Pruning and care are exactly as for a glasshouse vine.

The paving slabs act as storage heaters and so the Curate's Vinery is well able to ripen early varieties of dessert grapes.

The Vinery has only a very small volume and will need to be open on even the most slightly sunny day from Spring onwards and well open whilst the vine is resting.

The bunches of grapes should be suspended and off the paving… or else they will cook!

Vines in Pots

This is a good way to grow grapes for the patio or small conservatory.

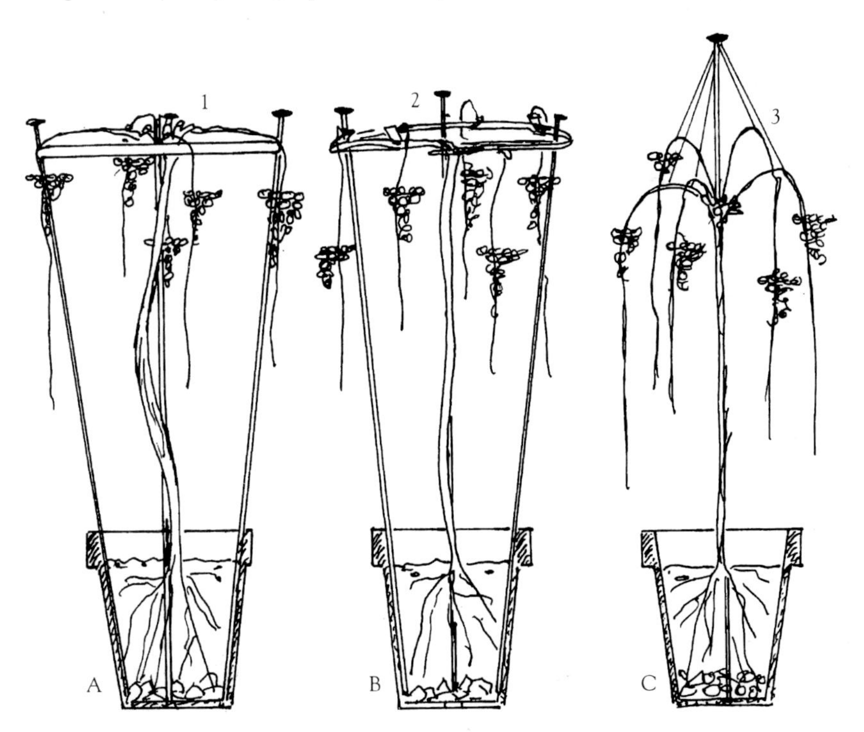

Pot Grown Vines

A. Spur pruned head.
B. Permanent spur pruned rod around a 'wheel'.
C. Spur pruned head.

1. Fruiting laterals trailed over a 'wheel'.
2. Laterals.
3. Fruiting laterals centrally supported.

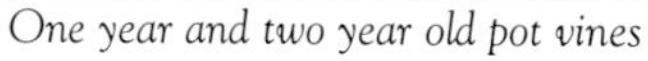

One year and two year old pot vines

Two year old pot vines...
spur pruned and permanent rod

The vines are portable and may be sunk outside for their Winter rest and then arranged along a sunny wall or on the patio before bringing them in to the conservatory or glasshouse for finishing.

The vine should be grown in a 5" (12.5cm) pot and grown on in a 7" (17.5cm) pot.

Its final growing pot should not be less than 9" (23cm)... A bigger pot is better, but you must consider the weight which you will have to move about.

Clay pots have the great advantage of stability.

Peat based composts are quite useless for vines and the best medium is John Innes No. 3.

The aim is to grow a Standard some 4' (1.2m) high and this is achieved in the same way as has been described.

The Standard is encouraged to grow five or six shoots which are eventually a spur pruned head.

Construct a supporting frame for the crop when the vine is in its final pot.

Some growers use one cane, to which the leg and the five or six fruiting spurs are tied, but I prefer this more substantial structure...

Insert three or four three foot six to four foot canes around the pot and push them right to the bottom of the pot and against the side.

Cut a five foot length of half inch polypropylene hot water pipe. Join with copper pipe (as for Trier Wheel) and drill the underside. Push the canes home.

The fruiting laterals arching from the spurs will be trained over the 'wheel' and then fall.

Alternatively the vine may be trained around the circumference and fruited at one bunch per foot of wheel.

Buckland Sweetwater is said to be 'easy to grow' and is often recommended for pot culture but in my opinion, whether easy or not it is the most bland of grapes and not worth the effort. Black Hamburg is much better flavoured and so are the Chasselas. The relatively new Dutch Sweetwater variety 'Roem van Boskoop' has good qualities. Indeed most of the early Sweetwater varieties are worth trying, but do restrict the crop to five or six bunches.

Feed with a high Potash fertiliser such as Maxicrop 'Tomato' every week until fruit set.

Ornamental Vines

It is part of a vine's genetic instructions that it should grow to occupy the greatest space possible and so the training of ornamental vines is solely a matter of confining them to the chosen site.

Just as with other vines, when cut too late, ornamental varieties will bleed from old wood, but green growth can be attended to with secateurs or shears.

The first year's cane should be reduced by two thirds but thereafter pruning is purely a matter of restricting the vine in whichever way you will.

Vines grow rapidly when once they are established and unless controlled a mature specimen can swallow a small shed with a season's growth.

Ornamental vines have mostly mauve, red and orange tints and so go well with other 'green' climbers such as ivies or they may be allowed to scramble through a tree.

PROPAGATION

Cuttings

Vine cuttings can be struck as readily as can blackcurrants and can increase stocks quickly in phylloxera free countries.

It is essential that the wood is hard and ripe and the best time to take cuttings is at pruning time and using the ripest wood from the fruited laterals.

Either tear the laterals off the discarded arm, with a heel and trim off to 9-12" (30cm) or make 12" cuttings from just below to just above a bud. A heel includes wood from around last year's bud and roots with ease. There is no point in attempting to root wood which is less than 'pencil' thick.

These cuttings are lined out in a sandy trench and planted with one or two buds above ground. I do this immediately after pruning but in cold areas it is perhaps advisable to plant them in bundles and line out in February. Alternatively 6" (15cm) cuttings can be rooted in pots and under cover where rooting will take place more quickly... A useful method when stock is in short supply. A gentle bottom heat helps and confining the potted cuttings in a polythene bag will help to keep a buoyant atmosphere around them.

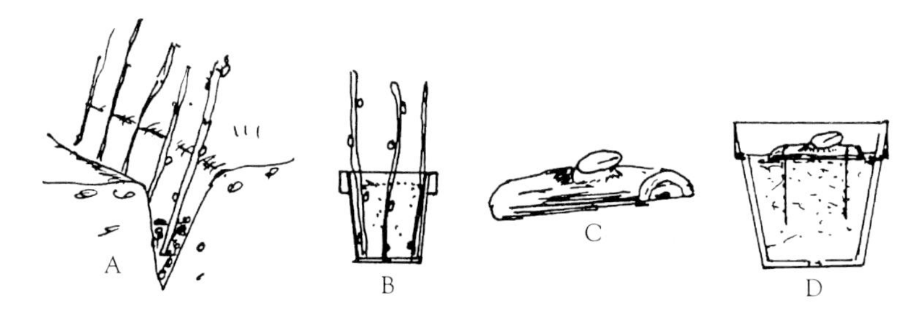

Rooting Vine Cuttings

A. Ripe cuttings in a sharp sand filled trench.

B. In a pot.

C. Prepared single bud cutting .

D. Anchored onto compost.

Vine Eyes

When only very small stocks are available, or in commerce, ripe wood is cut into sections, each of which has one bud in the middle.

The side opposite to the bud is scaped or pared away and then this short cutting anchored into sandy potting compost with about one third showing. One vine eye per small pot allows for planting out with minimum root disturbance. It is best to give gentle bottom heat.

Layering

This has the advantage that rooting occurs whilst the new vines are still attached to their parent.

It is achieved by burying all of, or a length of, cane about 2cm below the ground and anchoring with stones. The new plant(s) are cut away when the furthest reaches about 1" in height.

Grafting

This method is used for... grafting a European variety onto a phylloxera resistant stock... grafting a preferred variety onto a vine found to be inferior... or for grafting a good but shy setter of fruit onto a better setting stock... (e.g. Muscat Hamburg/Gros Colmar.)

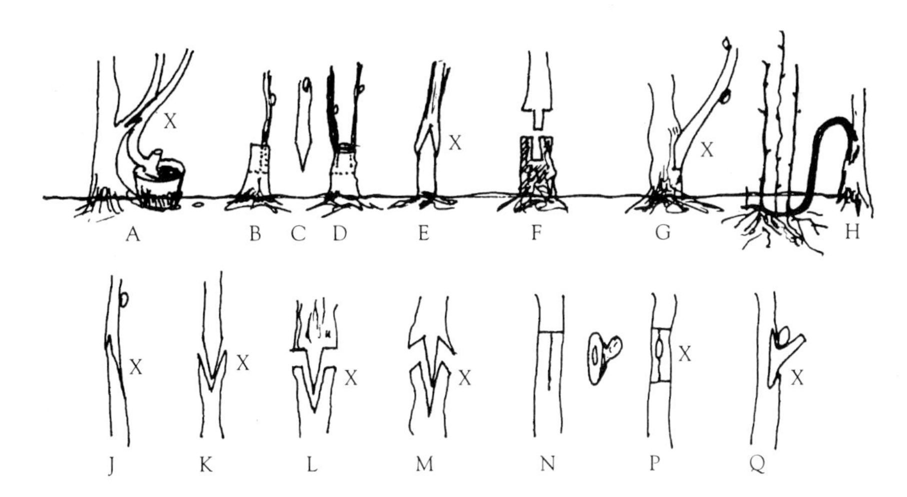

Grafting and Layering

A. Inarched grafting.
B. Simple crown.
C. Prepared scion.
D. Double scion.
E. Saddle.
F. Mortice.
G. Oblique.
H. Layer.
J. English Whip and Tongue.
K. Plain cleft.
L. Shouldered cleft.
M. Angled shouldered cleft.
N. & P. Bark cuts and bud and graft.
Q. Majorca bud graft.
X. To be grafting-waxed and raffia bound.

Grafting must be achieved quickly, for the wood dries as it is cut. The stock is cut first, then the scion and the two brought into close contact, bound with raffia and waxed as quickly as possible, for if any air bubbles enter the wound the graft will fail.

Grafting is a skill which some, never gain... and I am one of them! Others... including, fortunately, my wife, do.

The French bench grafting gangs of the 19th Century used the 'greffe anglais' (English whip and tongue) and laid the unbound grafts onto wet moss and sawdust kept at 25 to 30 Celsius. After a few days these grafted stocks could support 5kg (!) and were then rooted... It was these amazingly skilled men and women who saved the French Wine Industry, producing more than eleven million phylloxera free vines in quick time!

Modern bench grafts are most usually machine-made 'Omega' grafts. In effect this is a modified Shoulder graft. The stock and scion are cut and joined together by the machine which makes an inverted omega shaped graft.

A number of common vine grafting methods are illustrated. All are of use, but the 'English whip and tongue' and the 'Cleft' are the most used by gardeners.

The Bench grafting of scion onto rooting stock is best done at pruning time and the successful grafts can then be struck with bottom heat or heeled-in in bundles until the Spring.

Cleft grafting allows, for example, the grafting-on of an earlier ripening variety and may be lightly cropped after only a year.

It is done in early Spring as the sap begins to rise.

The stock is severed near to ground level and a knife forced down about 2". The freshly prepared scion, some 3-4" long is pushed into the slit with as much cambium in contact as is possible and the whole bound tightly with tight raffia and earthed over.

Whip and Tongue is also a Spring operation, but in this case the graft is bound and waxed.

Grafting is a specialist craft and I recommend making reference to a specialist source.

...Wine is sunlight, held together with water. (Galileo 1564-1642)

Chapter Three Summary

- ★ Prepare the site well and well in advance.
- ★ Buy your vines from a reliable specialist.
- ★ Buy grafted vines or maintain isolation.
- ★ Always address the vine along with its immediate neighbours.
- ★ Have a clear idea of what you wish to do BEFORE you use secateurs.
- ★ Deal with next year's needs before attending to the potential crop.
- ★ Use two differently coloured ties for the differently aged woods.
- ★ Be sure that a bud **is** blind before rubbing it out.
- ★ Next year's fruiting shoots arise from wood grown THIS year and so do remember to grow new spurs each year.
- ★ Remove unproductive growth as soon as it appears, remembering that a leaf guards the bud in its axil.
- ★ Always err towards undercropping your vines… take a maximum of two bunches of grapes per rod from a mature outdoor vine… or… one bunch per foot (30cm) of permanent rod from a mature vinehouse vine.
- ★ Grow for quality rather than quantity and expect little from your first crop.

Chapter Four

Feeding, Mulching and Maintaining Fertility

The General Picture

The label on a bottle of varietal wine from a single field or district will include its origin but, more often than not, the name recalls what the source was, rather than what it is.

For examples, Rosenhang (A rosey slope); Lay (A slatehill); Römerberg (A Roman Hill Cemetery); Bernkasteller Doktor was so named because a 14th Century Elector of Trier swore that it was its wine which had saved his life!; the wine from Senheimer Vogteiberg was once paid as tribute to the Elector's Vogtei (Magistrate).

Lafitte, Mouton (Mothon) and Tertre are Old French or Patois words for Mound... so importantly of gravel in Bordeaux. Valdepeñas means 'Valley of stones'; Nuit St. George's 'Les Vaucrains' means 'Good for Nothing' (Les vaut rien) and Wachenheimer 'Gerumpel' was formerly Wachenheim's rubbish dump. Very often no other crop could be grown, because of the lack of 'soil' or the precipitous nature of the site and it may be that it is consequential that the vine is often thought of as preferring to grow on lean land.

Nothing could be further from the truth!

The vine is a very deep and greedy feeder... it will take whatever there is to be had, whether it be in the ever moist gravels of Bordeaux, or in the reservoir from Winter snow in a Mosel Slatehill.

The vine will seek out the mineral solutions which are unique to the site – the extract – and which, with the essential factor sunshine, become the particularly luscious dessert grapes or the distinctively fine wine of that provenance.

Not all vineyards are favoured with apparently unlimited nutrients and they have all spent pressings and prunings returned together with any other organic mulches which can be found.

The fact is that most vineyards require appropriate and regular feeding.

MAINTAINING FERTILITY

In the Open Vineyard or on a Wall

It is a fortunate grape grower whose land lies over gravel or permeable rock or limestone, for these are sponges for nutrients.

Gardeners in East Anglia are on a Glacial dump of various clays, flint, chalk, shingle and stones some two hundred feet (70m) above the bedrock whilst those in Cornwall may be only a few inches from granite, but whatever is the geology of a garden it can be assumed that in The British Isles it is essential to muck or compost, to mulch and to feed vines.

Some gardeners believe that it is also necessary to irrigate, but I have never found this to be so, out of doors.

In my opinion, having been given a reasonable site and having been well mulched, it is then in the vine's best interests that it develops an extensive and deeply searching root system.

Each gardener makes decisions upon how to tend plants. I garden as 'naturally' as I can, being neither a fundamentalist 'organic' nor 'inorganic' grower. Good husbandry is, surely, a matter of ensuring that land is passed on in better condition than when it came and so must involve more 'give' than 'take'.

Prairies of unrotated monoculture drenched in 'cheap' Nitrogen and little else will be paid for by posterity whilst traditionally 'good' practices will leave land with a legacy of 'heart'.

When viewed at the points of delivery – the root hairs, where the uptake is as salts in solution – or the leaf stomata, where it is as Carbon Dioxide or sprayed nutrient solution – it becomes very difficult to argue that the nutrients enter the plant as other than anions, cathions or as Carbon Dioxide gas. There is no choice open to a plant but to use atmospheric Carbon Dioxide, which, although chemically 'organic', comes not only from organic sources but is manufactured by, for examples, motor vehicles, aircraft, industry and domestic fireplaces... and there can be no doubt that, chemically, the components of soil and subsoil are predominantly inorganic.

That a plant's diet is wholesome and balanced should be the main concerns, for there can be no uniquely correct way to cultivate.

The vine needs a good start and should be planted in clean ground into which has been worked plenty of organic material and 3oz per square yard (90g/m sq.) of John Innes Base or National Growmore.

Perfect sites, of which there there are few, may go for years without added sustenance but most usually the vineyard will have an annual requirement of about one ounce per yard run (30g/m) of John Innes or Growmore and as it matures the grapes and the wines will show the particular subtleties resulting from the vines deeper and deeper extraction.

The vine's leaves are green... if they become in any way yellow then all is not well. At worst you chose wrongly... a lime-hater or a lime-lover in a 'hated' soil... or as is the more likely, the leaves are indicating some curable deficiency... for examples, a yellowing between the veins indicates Magnesium deficiency whilst a mottled yellowing at the leaf margins shows a lack of Boron.

Quintessentially, healthy vines grow in healthy land.

Nitrogen, Manure and Compost

Manure and compost are fine providers of abundant humus and uncertain amounts and variety of trace elements. The Author's vines have lavish applications of well rotted strawy pig manure or household compost every second year, which will have a typical Nitrogen content of around 0.5%.

Manure is of unreliable availability and so I make compost each year. This compost includes all organic household and garden residues... fish, fowl, flesh, bone, 'hoover-dust', newsprint, tealeaves, weeds, prunings... whatever. (The bins are metal-lathed above and below for vermin proofing). All branch wood which is too thin for chipping or for the woodburner is 'Alko-ed' and composted together with all the wood ash from the household woodburner. The compost is turned only once, when Russian Comfrey is incorporated. This comfrey grows on thin poor land but gets Growmore and manure. Middling wood is chipped and spread between the rows of vines to make mulched and weedfree pathways and it is to redress the Nitrogen 'stolen' by the chippings that I use Growmore (7:7:7 NPK).

Foliar sprays of a Seaweed Tomato Fertiliser, typically 5% Nitrogen, can attend to immediate needs without inducing lush growth.

Lime (Calcium)

It is seldom necessary to lime, with the exceptions of the pH being below 6.4 or when growing a lime-loving variety. John Innes Seed and Potting Composts have adequate Calcium content.

Potash

This is the 'quality' element in fruit or grain of any kind and it is essential for the ripening of wood.

It is also the 'expensive' element.

Flabby, tasteless 'commercial' tomatoes and other fruits have resulted from crops pumped up with Nitrogen whilst being allowed to become deficient in Potash.

To have tasteful fruit and to ripen its wood the vine must have plenty of accessible Potassium.

Potassium has also the effect of hardening a fruit's skin which then resists the entry of fungus spores.

These needs may be satisfied with annual dressings of compost attenuated wood ash; with manure; with Growmore, or Potassium Sulphate, applied at an ounce per yard run at the end of February.

Phosphorus

Phosphorus is taken up by plants as phosphate ions and is indispensable in photosynthesis, when sunshine energises chlorophyll into combining 'breathed-in' Carbon Dioxide with water to make sugars and in root formation. It is found in dung and bone.

Suffolk's Compound Fertiliser industry began with the grinding and bagging of the Coprolite deposits from the Suffolk coast. It is commemorated as 'Coprolite Street' in Ipswich.

Coprolite which is very rich in Calcium Phosphate, is fossilised animal dung... organic or inorganic?

Phosphorus is now most usually applied as Superphosphate, which is produced from steel slag, or when further treated with acid to release more soluble phosphate, as Triplesuperphosphate.

The phosphorus in John Innes Base is as Superphosphate.

Magnesium

This is essential for chlorophyll formation. It is needed in smaller amounts than Nitrogen, Phosphorus and Potassium but as more than a 'trace' element.

Its salts are very soluble and so leach easily. One ounce per square yard (30g/m.sq.) of Epsom Salts, worked in prior to planting, will suffice for the first year with subsequent annual Spring dressings of half an ounce per yard run or the Epsom Salts may be given together with one of the foliar fungicide sprayings. Two ounces in a 3-gallon knapsack sprayer applied as a spray to run-off is plenty to correct a seen deficiency..

Alternatively the Magnesium level can be maintained with an application of Kieserite (Commercial grade Epsom Salts), outside or under cover, at about two ounces per yard square (60g/m.sq.), every second year.

Boron

Millerandage or 'Hen and Chickens', when the bunches set unevenly and with variously sized berries, is caused by Boron deficiency. The vines make shoots unevenly and the growth is brittle.

A great deficiency results in small and misshapen leaves and die-back.

Similar effects are seen in all fruiting trees and bushes and root crops become stunted and necrosed.

It is essential as a trace element, but in excess it can be fatally toxic.

One sixth of an ounce (5g), about a teaspoonful of borax, in a 3-gallon knapsack sprayed without run-off is a safe application for one year's needs. The best time to spray is at the three leaf stage and before flowering.

Borax is available as 'Solubor' commercially or as 'Borax' from the Chemists.

Seaweed Extract contains around 40mg/kg, of Boron and so a deficiency is unlikely if a 'Maxicrop' programme is practised.

Zinc and Manganese

The availability of these two elements is critical, particularly throughout flowering and pollination.

The easiest way to give them is as a Dithane 945 spray – the commonly used potato blight preventative – for in so doing the vine is also protected against fungal attack.

Seaweed extract is another reliable source for both metals.

The spray programme should begin when the leaves are about 50p piece sized and a couple more times until flowering begins and should recommence after fruit-set and fortnightly thereafter.

Coulure, the sudden necrosis of flowering shoots, may be caused by Boron, Zinc and Potash deficiency or combinations of lesser deficiencies of two or three of these trace elements and, very occasionally, together with lime deficiency.

Coulure has a further hazard in being a focus for botrytis.

Shanking, is the most disappointing happening for the grower, coming after the bunches have been selected and the crop begins to show promise. The first sign is a brown spot on the berry's stem which eventually it encircles. The grape ceases to receive sustenance and is useless. Whilst not fungal in itself it opens the bunch to fungal attack and so must be removed. It is, in effect, "late coulure" and is the vine's way of trying to make insufficient nutrient, sufficient… it recognises being overcropped and underfed… Too many bunches can also initiate the problem on an otherwise healthy vine, but both coulure and shanking are due, usually, to chronic bad practice, which sometimes dates from the preparation of the ground, for the vine may unable to grow a sufficiently efficient root system because it sits in a too wet, or too dry and infertile an environment. In the short term, beginning a foliar spray programme may save the rest of the crop, but a long term improvement will result only from improved practice.

Its prevalence may lie in the misapprehensions that grapes should not be fed and prefer to be in poor ground. In truth, whilst vines **will** become lush and prone to disease if they get too much Nitrogen, they require high levels of other plantfoods. It may be necessary to replace the top soil around the vine and aerate the subsoil, but in any event the vine must be mucked or given compost and treated as described earlier in this chapter.

Iron

This is seldom deficient, but deficiency shows as stunted yellowed growth and eventually death.

It is more often the case that a lime hater is prevented from taking up Iron by Calcium.

It is unlikely to concern the grape-grower who grows suitable varieties but the easiest correction is made with sequestered Iron.

Other Trace Elements

Many other elements are required by plants in minuscule quantities. Every living thing on the Natural Carousel is what it ate... what went around, comes round!... and so, if a variety of natural residues is used as part of a fertiliser programme, for examples... Manure, Compost, Hoof and Horn, Blood, Fish and Bone, Shoddy or Seaweed, and the vine has an extensive root system, then it is most unlikely that its tiniest trace needs will be neglected.

The Author includes Seaweed extract in every spray and believes that it attends, not only to the supply of many trace elements directly to the leaves, but also to making the whole vine less palatable to pests.

Under Cover

Maintaining fertility in the vinehouse is very like maintaining fertility outside, with the exceptions that since bigger crops are taken, generally heavier applications of fertiliser are required and because it neither rains nor snows inside, it will be necessary to irrigate for part of the season.

My vines are planted along the inside of the vinehouse but I prefer to muck and compost along the outside and the programme begins in late January, with an application of one once per yard run (30g/m) of Potassium sulphate, outside the vinehouse.

During February two ounces per yard run of National Growmore or John Innes Base goes down along the vines.

A further half to one ounce per yard run may be applied after the fruit has set.

From when the leaves are 50p sized and whenever the vines are sprayed for zinc needs, or against insect or red spider or for fungus resistance, include a spray to the manufacturers' instructions... for examples, Maxicrop Triple Biostimulant, which is the Manufacturers' recommendation, or Maxicrop Tomato Fertiliser (NPK 5.1, 5.1, 6.7) if an extra need is evident. (Typical Analyses of common fertilisers are listed later.)

The house should be watered sufficiently and kept buoyant. It and the vines should be sprayed daily from budburst until set, after which the vines must not be sprayed. Watering must cease at veraison (colouring of the grapes).

Whilst it is preferable to muck and compost outside the house and so promote their weathering, the organic content of the borders' soil should be enhanced, every second year, with some relatively sterile humus provider... for examples, peat, cocoashell, feathers or shoddy.

Vines in Pots

During the first year there will be sufficient nutrients in the J. I. No. 3 but there is purpose in spraying regularly with Seaweed extract.

At the end of the second season and all subsequent seasons the top 25mm of the compost should be replaced with fresh J. I. No. 3.

Apart requiring this soil replacement, pot-vines have the same needs as do other vines.

...Wine brings to light the hidden secrets of the soul. (Horace 65-8 B.C.)

Chapter Four Summary

★ Plant into well prepared land with plenty of organic material.

★ Keep your land in good heart and as 'natural' as is possible with regular applications of compost and muck.

★ Be mean with Nitrogen and generous with Potash.

★ Attend to trace element needs. For example… Zinc and Manganese are essential for successful setting.

★ Vines under glass crop more heavily than those outdoors and so need more food and will need irrigation.

★ It is preferable to spread muck and compost OUTSIDE the house and dress INSIDE with fertilisers.

★ Use John Innes No. 3 for pot growing vines.

★ Seaweed is a sovereign plant food.

CHAPTER FIVE

Pests, Disease and a Spraying Programme

There are many vine pests and diseases, most of which and including the worst, we are free from in the United Kingdom. The most feared is the aphid **Phylloxera vastatrix**. At present we are phylloxera free and it is sufficient to be able to recognise the pest and get on with the enjoyment of viniculture, for, if your vines are infested with phylloxera there is no saving them and the pest is notifiable.

Problems with pests come, most usually, from… Red spider mite, Glasshouse whitefly, Scale insect, Mealy bug, Vine weevil, Wasps and Birds.

Occasionally greenfly migrate from weeds, a family of earwigs grazes some young leaves, there is an experimental nibble from a capsid bug or some leaves are colonised by Erineum mite. These latter mentioned are usually hardly worth a spray and may be dealt with by hand.

Glasshouse Red Spider Mite

If what you see is red and/or a spider then that is what it is and it will be harmless or beneficial.

Red spider mite is tiny and brownish to blackish and infests the undersides of leaves.

The upper sides of infected leaves become silvery yellow and, if the infestation is severe and unchecked, the leaves and eventually the plant will die. As its numbers increase the mite spins a dense web.

It enjoys hot dry conditions and it is resistant to all but a few chemicals which are also hazardous to Man.

The best control is to be without it. It dislikes moisture and it is unhappy in the buoyant atmosphere which comes with daily spraying. An infestation may be held in check by a daily spraying, with force, of the undersides of the leaves.

If it has become rampant then Malathion or Rogor at fortnightly intervals and daily spraying will recover control.

Alternatively and preferably, Phytoseiulus mites, red spider mite predators, may be introduced when their optimum breeding temperature can be held. Quite obviously the use of predators precludes the use of pesticides in the glasshouse.

It may be, or it may not be the case, but I believe that Seaweed extract makes greenhouse plants less palatable to Red spider mite.

Fruit tree spider is seldom of consequence on out of doors vines.

Glasshouse Whitefly

The adults are tiny moth-like creatures which drift around when disturbed. They infest the undersides of leaves where, their nymphs, in particular, do considerable damage. The honeydew, which they excrete, encourages the growth of sooty moulds. They are resistant to most, if not all, insecticides and the best control is kept with the Encarsia wasp which is introduced during the Summer. Like red spider mite they dislike daily spraying. A cordless vacuum cleaner will dispose of disturbed and thus airborne whitefly.

Scale Insect

This is an insect which the vine shares with Cacti and other greenhouse plants. It is first noticed as a fixed browny grey domed scale about 6mm long and roughly oval in shape. The insect is not particularly mobile in any part of its life cycle and will have been introduced to your greenhouse with the vine or on some other plant... Look carefully at any new purchase or gift.

Under the scale is the a female, most usually dead, and a great number of eggs.

The odd one can be picked off and the eggs smeared.

A December Tar Oil Winter wash will dispose of this and other pests... and, unfortunately, beneficial creatures too.

It is not found on outdoor vines in the U.K., because it cannot complete its life cycle.

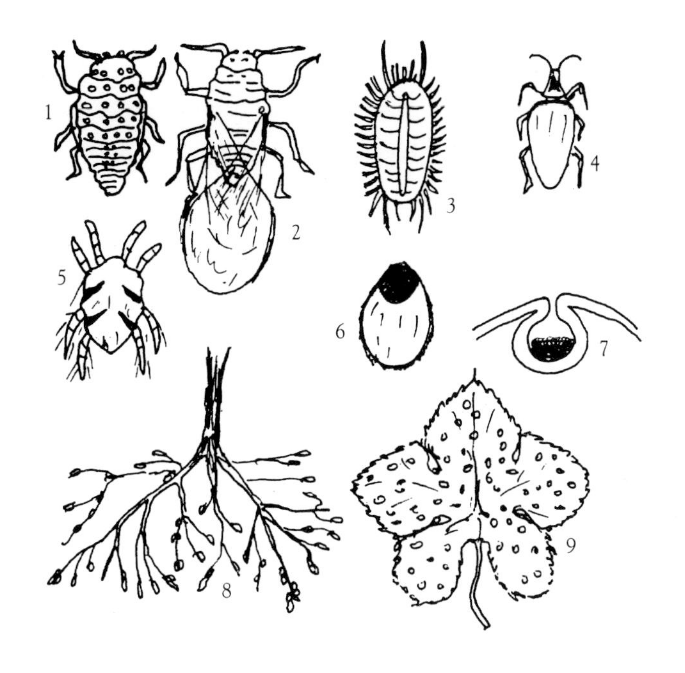

Vine Pests

1. P. vastatrix wingless female.
2. P. vastatrix winged female.
3. Mealy bug.
4. Vine weevil.
5. Red spider mite.
6. Scale insect.
7. Section through a root gall with eggs.
8. Vine roots infested with Phylloxera nodosites.
9. Vine-leaf (underside) with galls made by wingless females.

Mealy Bug

Any attack is usually from Cactus mealy bug, which is not native but is common on greenhouse plants, or from the native mealy bugs, most often seen on meadowgrass and sometimes on lettuce roots in dry Summers.

A weedfree environment discourages and two Rogor or Malathion root drenches a fortnight apart or the introduction of the Ladybird – like Cryptolaemus destroys the pest. It is seldom of great consequence.

Vine Weevil

This was the pest of the Victorian vinery but, in more recent times, it has become common out-of-doors, most particularly on rough and neglected ground.

It has a catholic taste but it is most comfortable on any pot plant or vine.

It has enjoyed a population explosion over the past few years, due in the main to the use of peat based, soilless potting composts.

It is damaging to the vine as a dark grey adult, with its nocturnal grazing of leaves from the margins inwards and as a 12mm brown headed, fat, cream coloured grub, when it destroys the roots. It is killed by GammaBHC, which is also harmful to Man and it may be damaged by Derris dust and French Chalk, which are not.

The slow moving adults can be caught and destroyed with the help of a torch in the dead of night.

Predator Nematodes are available.

Vine weevil is unlikely to debilitate a mature vine but it can cripple a pot grown vine. Nematodes will give excellent control in pots.

Wasps

These are a problem outside only in an early year on an early variety and in an extended warm Autumn but they can be of great annoyance in the greenhouse.

The only really satisfactory control is to find and destroy the nests.

If you happen to discover a nest during the Summer, mark its situation but do not destroy it (Unless it is in the house!), for in the Summer time the wasp is most beneficial. Wasp larva need protein, which is brought to the nest as greenfly. Their 'pest' stage comes when the brood is complete and the insects need only sugar to thrive.

'Nippon', which is syrup and borax or, more cheaply and selectively, 'Best Bitter' and water and borax will poison wasps and, if they get back to the nest, the fed queen.

'Best Bitter' is not attractive to bees but sweet substances are.

Grapes grown in polytunnels are easily protected by dropping the net ends. Similarly, nylon netting over the ventilators will protect the crop, but this is an unlikely method for any other than a small greenhouse. I find that individually tailored bags made from fine netting give total protection.

Birds

'Buzzline' deters but the only guaranteed protection comes with netting.

My vines share a permanent fruit cage with gooseberries and currants but commercial growers draw a net over the vines immediately before harvest. Wall grapes are easily protected by draping a net over them.

I have heard that a popular method on the Continent, although I expect not with neighbouring vineyards, is to wait until the starlings have just settled in the late evening and then to fire shot-guns over the vineyard. The startled birds take to the air and settle elsewhere for tomorrow's feast!

Other Pests

Very occasionally a family of earwigs will nibble holes into a young leaf, which makes it unsightly and of reduced use as the holes grow with the leaf. It is not at all like vine weevil damage, which is from around the edges of the leaves. Earwigs are fine scavengers and do far too much good to be harassed! If surprised they will fall to the ground and run off, quite probably **not** to another vine.

Capsids sometimes try a vineleaf or you for taste, but do very little damage.

Erineum mites may colonise a leaf or two. They live under the leaves and cause blisters on the upper sides. The damage is quite unlike Red spider mite damage which 'silvers' leaves. If the leaf is turned into compost… then so are the mites.

General Pest Control

There should be no need for any systematic chemical control of vine pests.

Cleanliness and the maintenance of a healthy environment are the main pest deterrents, helped I believe and with some scientific evidence, by regular sprays of seaweed extract. If there is a sudden and alarming infestation by aphids, mealy bug or red spider mite then two applications as a root drench of systemic pesticide will lessen or eradicate the problem without stressing predators too much, since they prefer to hunt live prey.

DISEASES

Oidium… 'Powdery Mildew' and Perenospera… 'Downy Mildew'

It makes more sense to attempt prevention of these rapidly spreading fungal diseases than to try to cure them.

Oidium tuckeri begins as dark grey patches of mould on the leaves and fruit. Eventually bunches and leaves are covered with a grey powdery coating. The fruit becomes, hard and cracked and ceases to ripen. Not only is the fruit useless but it will taint wine.

Perenospera viticola begins as oily looking patches on the leaves. If it is not checked it destroys the leaves and so the crop.

Control Out Of Doors

A fortnightly spraying with Seaweed extract will toughen the foliage and so, resist infection.

The inclusion, alternately, of wettable Sulphur or wettable Sulphur and Dithane will give good defence against both mildews. Mancozeb and Zineb are similar compounds to Dithane and may be more easily obtained, or be cheaper in some areas. Sulphur, which is traditional in its usage and so 'Organic grower' tolerated, has been used for more than a century without mildew building up any resistance. It is most efficacious after the heat of the day. If wettable Sulphur is unobtainable then Flowers of Sulphur may be made 'wettable' by mixing it with a small amount of washing-up liquid before admixing it into the sprayer… not a lot or else the spray will foam. Sulphur is most easily sprayed using a coarse nozzle. Dithane and related compounds contain the trace elements Zinc and Manganese, both of which are essential in fruit formation. They give excellent control of Perenospera.

Both mildews may be controlled with Copper and Sulphur, in the form of "Bordeaux Compound', which is traditionally 'organic' but I do not recommend the use regular use of large amounts of copper, for it is cumulatively poisonous and each application sets the crop back a week or ten days. My German Winemaster friend never uses Copper. If it is to be used then the best time is immediately AFTER harvest and once only each year, when it cannot be absorbed by the vine and has the effects of hardening growth, disinfecting and protecting against, not only mildew and botrytis but also… **Phomopsis… 'Dead Arm'** which identifies itself in the Spring as white, dead looking… and they are… replacement canes or cordons. They must be removed and burned. Vines are opened to attack from other fungi by… **Botrytis… Grey Mould**, the ubiquitous destroyer of fruit of all kinds. There are a great number of chemicals which are said to control botrytis, but if they do then it is not for very long for the fungus mutates to resistant strains with alacrity. If artificial chemical control is to be practised then more than one chemical must be used and the programme altered before the fungus begins to show resistance.

Sulphur, alone, maintains some control without harming Man, but at least one other of the chemicals has been associated with birth defects in humans.

Preventative spraying should begin about ten days before budburst but not during flowering and fortnightly thereafter until at least a month before harvest, because, whilst Sulphur will be precipitated after milling grapes for wine, other chemical residues may not be and can retard fermentation and cause taint.

Control Under Cover

Again, preventative spraying begins about ten days before budburst, but whilst Seaweed extract is used fortnightly, Sulphur and Dithane are added to the spray together and only monthly. This is because the fungicides are unlikely to be washed off under cover. They should not be used within a month of fermenting or eating.

Sulphur is more phytotoxic under cover and so it must be applied only in a damp atmosphere and in the late evening.

Winter Wash

A 2% Tar Oil wash with Copper and Dithane under glass, not later than the end of December and not later than late January out of doors and will benefit vines more than they lose.

Once every other year is often enough out of doors.

Some growers do not use copper and some will leave the Copper and/or Dithane spray until March or April. Spraying must not be carried out during, or with the threat of, frost.

Recommended Spray Mixtures

NOTE: All Mixtures include Seaweed Extract and assume the use of a 3-gallon sprayer.

Pre-blossom

2oz (60g) Sulphur, Dithane to Manufacturers' recommendation, 1oz (30g) Epsom salts, 1/2oz (15g) Borax.

Post-blossom

Alternatively... 60g Sulphur or 60g Sulphur and Dithane when the fruit has well set and not before and, fortnightly out of doors. The bunches and the undersides of leaves must be drenched.

The programme will be based on the expected time of harvest so that 'the last spray' is as close to a month before picking as can be, 'The last spray' should be as for pre-blossom but without the Borax.

Some help with regard to a spraying programme may be gained from...

Approximate Times to Ripening

Madeleine Silvaner is the earliest and ripens about 60 days after setting.

Siegerrebe needs at least 65 days to ripen; Madeleine Angevine needs at least 70 days; Müller-Thurgau at least 90 and Riesling is the latest needing at least 110 days under ideal conditions.

Virus Diseases

Leaf Roll Virus is not transmitted to other vines. It is evident in Summer when the leaves become yellow and roll inwards. It should not be confused with Autumn colouring. When confirmed the grape should be removed and burnt. Replanting may be immediate.

Fan Leaf Virus is serious and is highly transmittable. The leaves are undersized and crinkly fan-shaped and the growth between nodes is zig-zag. During the near past, when MCPA and other hormone weedkillers were cheap and much used by farmers, spraydrift caused damage which was very much like Fan leaf virus but from which the vines generally survived to grow properly the next year.

Any vine confirmed to have Fan leaf must be burned and its site left vacant for many years. It is a very uncommon virus and the symptoms may still generate from carelessly applied lawn weedkiller.

Entipiose Virus causes a vine to start poorly and its leaves to die. It is not yet a problem here but has become endemic in large parts of Europe. The vine must be burnt and the station left empty for three years.

Mosaic Virus is easily identified by one half of the leaf being smaller and yellow green. The vine must be burnt and the ground left unplanted for many years.

Conclusions

The list of pests and diseases is long, although nothing to the list faces growers in warmer climes.

Do not panic! The dreadful horror which is apparently afflicting your vines is far more likely to turn out to be minor and treatable. Buy good stock, and keep your vines weedfree and organically resistant and few problems will occur with which you cannot cope.

The hoe and mulch give the best weed control.

'Roundup' is excellent in ground preparation and for spot eradication of Bindweed and Ground Elder, but be warned that many professional growers of differing crops are suffering a loss of fertility on land which has been routinely weeded with herbicide.

...It is said that wasps dislike the scent of lavender.

Chapter Five Summary

★ Roundup is useful for clearing new land but do not over-use.

★ We are free from most vine pests, including the worst.

★ There should be no need for routine chemical pest control.

★ Daily spraying and Seaweed spraying reduce the likelihood of pest infestations.

★ Vine weevil appears to enjoy living in peat based composts.

★ Avoidance is better than treatment… Routinely fungicide spray with Sulphur, Dithane and Seaweed from 50p leaf stage to one month before harvest.

★ Use the hoe to keep down weeds and thus pests.

CHAPTER SIX

Harvesting, Preservation and Vinification

Veraison, the changing of the grape's colour from green, marks the **beginning** of the ripening time. The grapes are still remarkably hard and uneatably acid. It may be from four to six weeks before they are ripe and late Muscats and most Vinous varieties can need ten weeks to ripen. Many fine dessert grapes announce their readiness by releasing a most mouthwatering perfume but very often the approaching ripeness of Sweetwater varieties is heralded by impatient wasps.

In brief, ripening is the process of making sugar and losing acid. With a perfect balance of sugar and acid comes perfect ripeness. It is a process which is dependent upon sunshine and for the vine, a daily average temperature of, at least, forty eight degrees Fahrenheit (10°C).

The main players in ripening are... Malic acid ('Apple acid')... which is changed into... Tartaric acid ('Grape acid')... which is changed into... Fructose and Glucose (Fruit sugars). These are all carbohydrates which are related each to the other in structure. It is a wondrous miracle of nature that the loss of an atom of Oxygen here and a little splitting and rearranging there can change the inedible into the delicious.

Most dessert grapes finish with a very high ratio of sugar to acids. Tartaric is less acid on the palate than is malic. A ripe dessert or a ripe wine grape will have as little as one tenth of its acid as malic.

Ripening is a one way process and so in a poor year too little malic acid is changed into tartaric acid of which, then, there is less available and the result is odourless, thin, sharp and unbalanced wines. Whereas in a fine year such as 1976 which is said by German Winemasters to have been ... 'The best year since the Romans'... English dessert grapes were abundant and fine and many 'Great' wines were made in Continental Europe.

It is one of the advantages with growing under cover that, even in a coldhouse the ambient temperature can remain above an average of 48° Fahrenheit well

Veraison... Black Hamburg

Veraison... Mrs. Pince's Black Muscat

Edelfaule... Mosel, Germany, late in October

Noble Rot...Suffolk, England, early in November

into November… a warm Autumn can save a crop.

Harvesting Dessert Grapes

Do not handle the bunch or the 'bloom' will be spoiled and so will the appearance of the bunch.

When it is certain that the bunch is ripe it should be cut as a 'T' with lengths of lateral. This makes a convenient carrying handle and an umbilical for preserving Muscat and Vinous varieties.

Sweetwater varieties have thin skins and are to be eaten as they ripen. Of course they do not all ripen at the same time and you may be fortunate enough to be picking Black Hamburg in December.

Some Muscat and all Vinous varieties can be kept well into the next year in a 'Grape Cellar', which may be no more than a dark cupboard in a cool, but not

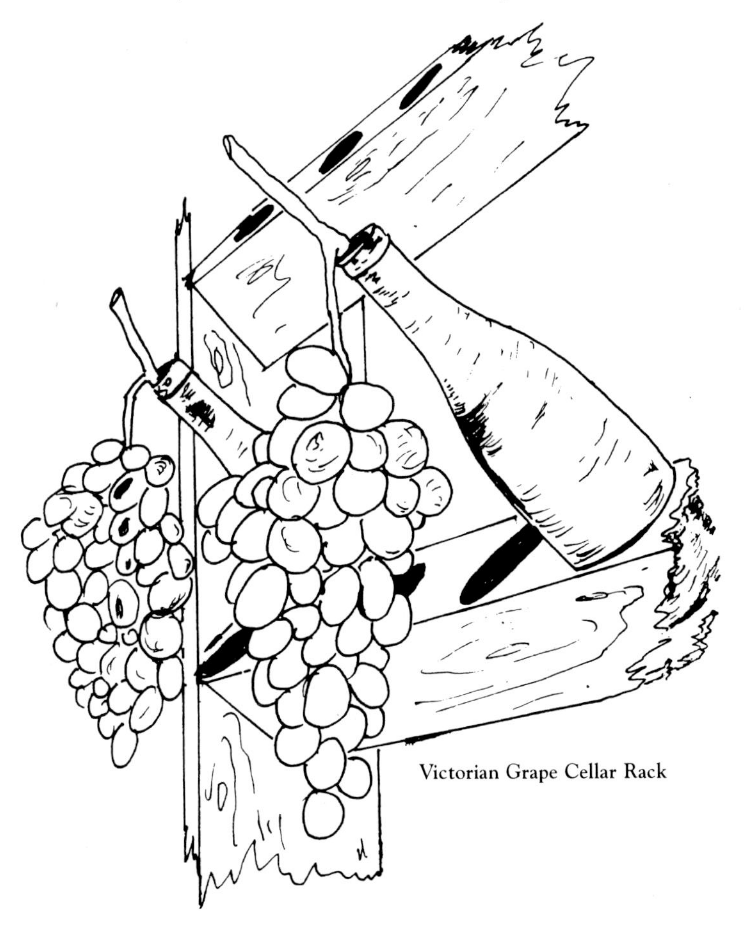

Victorian Grape Cellar Rack

very cold, part of the house.

A 'Victorian' Grape Cellar Rack is quite easily fashioned from ply and 6" x 1" planking.

'Victorian' grape bottles, which are sometimes 'discovered', or wine or milk bottles are canted at forty-five degrees and partly filled with water and kept fresh with a little charcoal. The bunches are cut with a "T" but with the arm nearer to the rod cut longer. It is the longer arm which goes into the water. Slightly unripe bunches from the cold house will continue to ripen if the temperature is kept above 48 degrees Fahrenheit. Temperate treatment is best... they will not 'cook' to ripeness!

Harvesting and Fermenting Wine Grapes

The Gardener Winemaker should aspire to producing wines which are drinkable and with a personal style.

Many 'Weekend Winemasters' in Germany, own or rent parcels of vineland and tend their plot, as a self-rewarding and self-supporting hobby. Some make wine which is good enough to be regarded highly by their neighbouring professional growers.

If a gardener grows suitable grapes and grows them well and then maintains good winery practice, then his or her wines will improve year on year and should become the equal and better than some on supermarket shelves... certainly different and that with having given the pleasure and the skills to bring it from earth to the bottle **and** the certain knowledge of its purity.

The Wine Harvest

Winegrapes, like Cider apples, have been bred to have the right qualities for fermenting when they are fully ripe. They are far more acid than dessert varieties but have more sugars too. They should be left on the vine for as long as the grower has the courage to leave them... As an East Anglian grower I have never had a crop with too much acid since the first, and the last five crops of Siegerrebe, Schönburger and Müller-Thurgau, have had acid weights of between 8.8 and 9.2g/litre before fermentation.

Initial acid levels of between 8 and 12g/litre are fine. If the acid content is more than 12g/litre it is the usual recommendation, but not mine, to add 'Acidex'

(Precipitated Chalk) at 3g/5 litres for each extra gram per litre of acid. I believe that if you are to know your grapes you must go along with them. Such a philosophy is not promoted by an attack on the grapes as soon as they are off the vine! My first vintage was pretty grim stuff, in the main because I picked too soon, but I learned... that picking too soon results in high acid, low sugar and a sharp and unbalanced wine with little body, but having made it myself, it was consumed and it was certainly an educational experience! A couple of the Gervin yeasts, which are available to amateurs, are able to metabolise 25-35% of the stronger malic acid into ethyl alcohol and Carbon Dioxide. Our 'Indian Summers' allow us a long ripening period, greatly to our advantage. Even in early to mid-November a couple of days can give a degree or so of sugar and whilst the vine functions it is extracting minerals. Long slow ripening gives a varietal wine nuances of nose and taste which it lacks when ripened quickly and hot. Fast, hot ripening results in high alcohol potential, but suppresses the take up of nutrient. Northern European white wines are famed, not for their inebriating power but for their complexity and charm. During the middle of September the leaves will become increasingly autumnal looking and they should be removed, gradually, to let in as much light and air as is possible to the grapes.

Now, the tongue is not a reliable tester for sugars, for the particular sweetness to the taste of a variety will be altered by its acidity. A high acid variety may be sweeter than a less acid variety which seems to be sweeter on the tongue.

The Hydrometer is a cheap and easy instrument to use for determining sugar content but it needs quite a lot of juice. The refractometer needs only one grape for testing, but it is a very expensive scientific instrument. I use a hydrometer but preserve the juice after testing in a screw topped bottle together with half a Campden tablet to be added to the must later.

A Specific Gravity of 1080 (80 degrees Oechsle) means that the juice has the capability to produce a wine with 10.6% alcohol and, in my opinion, most English white wines need no more, in fact I prefer 75-78 Oe (10% Alc). (Appendix 'Hydrometer. Use and Table').

Red wine needs to start from around 90-93 Oe (12 to 12.9% Alc).

My Suffolk experience has been that Gagarin Blue and Tereschkova, ripen outdoors to be pleasantly eatable, but do not finish sufficiently to make other than a thinnish 'pink'. A far better and pleasantly scented red comes from blending the outdoor harvest with greenhouse grown Gagarin Blue, Tereschkova and Dornfelder and which will then need little or no added sugar to reach 90 Oe or higher.

All three ripen much better on a warm wall under cover in the east but will succeed in the vineyard South of the Thames. If juice has a gravity of less than 55 Oe, then it will not make a worthwhile wine and the grapes should be left for later. It will have been a dreadful year, or you have the wrong variety, if the grapes cannot manage more than 55 Oe, by November!

When the decision to pick has been made then pick only as much at a time as can be processed immediately, for oxidation can be the ruination of a wine. 7kg (15lb) will result in 4-5 litres (about a gallon) of wine.

Hard mildewed bunches must not go into wine. The grapes have not ripened in any way and the mildew will taint.

But, if your bunches have become infected with botrytis **after** ripening, then Rejoice!... for they have 'Pourriture Noble'. 'Edelfaule'... the 'Noble Rot', the stuff of Sauternes, Beeren-Auslese and Tokay. So long as the berries are moist, as is a raisin, then, horrid and hairy though they may look, the wine will be improved by their inclusion. If there is plenty of noble rot then it should be reserved for 'Dessert' wine starting at around 120 Oechsle.

'Noble rot' should **not** go into Red wine or Pink, **only** into White.

Winemaking Equipment

You will have had three years to gather equipment and there will be another three before the vines near their cropping potential.

If you have been a home-winemaker then you have a start. You must have a Hydrometer. Either, with a cylinder, or better, with a Glass jacket and Bulb. You must have a press. There are a number on the market and your local Homewine shop can advise you to match your likely need. (Addendum... Home-made Wine press). You can, of course, tread your grapes, but it is less sticky if you mill them. An old rubber or wooden roller mangle will do the job if set horizontally and very small quantities can be put into a net bag and crushed by hand or with a pastry roller, or crushed in a household bucket with a wooden dolly.

My preference is for a mill. This can be had from a wine shop or quite easily home-made (Appendix... Grape mill). A suitable container catches the milled fruit. Five gallon (30 litre) fermentation buckets with fitting lids are ideal. Fermentation takes place in 5 litre (one gallon) demijohns or larger plastic fermenters with fermentation traps.

Syphoning tube with a lees excluder will be needed for racking and eventually you will require bottles and a corker (or access to one). If you are buying, then

A 'DIY' Grape-mill

A 'DIY' Wine Press

Sundry winemaking equipment

the French metal Lever Corker is the best. Plastic shrinkable or foil capsules improve the look of a bottle of wine and protect it from the attentions of cork moth.

They are sold, usually in 25's and in various colours.

Very useful extras include... a small cylinder of Carbon Dioxide for 'blanketing' the juice; 2 gallon household buckets; a largish funnel; some fine nylon netting to filter out 'grape bits' from the pressed juice. The only metals which can be in contact with wine without tainting are Stainless Steel and Aluminium.

Testing for Acidity is done easily with an inexpensive Kit and is much more reliably done than with indicator papers... we wish to know the **quantity** of acid (g/litre), not its **degree** of acidity (pH). Universal and narrow range Indicator papers are very inexpensive but give a less accurate picture of amount... pH 3.0 to 3.4 will mean that there is sufficient acidity for a good fermentation. A guide to sizes for equipment can be gained from the size of the vineyard... One vine produces from 3 to 5 pounds (1-2.5kg) of fruit and 15lb (7kg) of fruit can make around six bottles of wine.

FERMENTATION

The Starter

When the die is cast, the vintage begins with clean sterilised equipment. There is no need for clinical cleanliness but all containers should be washed with sterilant (e.g. Boots Sterilant and Cleaning Powder). Enough grapes are picked to make about a glassful of juice which is heated to 190°F (90°C) and held there for a couple of minutes to kill all wild yeasts and spoilage bacteria.

The juice goes into a covered bottle and when it has cooled to about 65°F the chosen yeast is added. Any wine yeast will do a good job but I use, always, a suitable 'Gervin' pure yeast. Gervin yeasts are very reliable and are used by a majority of professionals, throughout the European Union. This 'Starter' is plugged with cotton wool and set aside in a temperate place. Gervin 'B' (Strain CC) is an excellent yeast for making Rhine/Mosel type wines. It encourages a fine bouquet and has the capability for metabolising 20-25% of Malic acid into ethyl alcohol and Carbon Dioxide. Fermentation should be done at around 15 to 17 degrees Celsius (60-65°F) to maximise the production of 'fruity' esters, but Strain CC will ferment at 10°C.

Gervin D has similar qualities but is not happy at as low a temperature and is perhaps at its best as a cider yeast.

Picking, Milling and Pressing

The bunches should be picked into plastic buckets, weighed and recorded and milled as quickly as possible.

There are two schools of practice with regard to de-stalking and I belong to the one which does not. (With the exception of, for example, Gagarin Blue which has soft greenish stalks which taint wine).

The Mosel view, which is the one to which I subscribe, is that "Wine comes, not from grapes but from bunches of grapes… it is made not from 'juice' but from 'extract'". This is born out by the facts that, whilst grape juice has less than 0.05% Tannins, the stalks have 3% and the pips 5%. Tannins are essential for bite and character. There is the bonus, too, in that stalks open up the pulp and so help the flow of extract. Too much tannin makes wine astringent and so care must be taken to avoid abrading the stems or crushing the pips.

There is now a choice with white wine… to press immediately, or to leave the milled bunches for a few hours.

My practice is to add 'Pectolase' (an enzyme) which breaks down pectins and releases extra sugars and flavour, and to leave the pulp overnight before pressing early the next day. It is essential to keep air out, either with a sheet of polythene or with a blanket of Carbon Dioxide.

Pressing should not be overdone. The very best quality juice is the 'Free run' juice which runs without other than the pressure from the pulp's own weight. This can be set aside for a 'Premier wine' or not.

Pressure should be applied gradually and should not be continued after the time when only a little juice runs. The rule of diminishing returns applies and the later the juice, the thinner.

The pulp is emptied, rearranged and then returned to the press when a little pressure will cause some previously inner held juice to run. The spent pulp goes back to the vines or to the compost heap, but some amateur winemakers cover the spent pulp with water for 24 hours, excluding air, press again and Chaptalise to 70° Oe to make a very thin 'Country Wine' before re-cycling. Frankly, it is not worth the effort.

It is advisable to run the juice through a sieve and then a fine nylon net on its way to the bucket to catch pips and skin which has escaped the press. A little goes into the hydrometer jar and the rest into the fermentation vessel.

Some allow the extract to settle for a day or so and then rack into the fermentation vessel, but I have found this to be unnecessary.

If the must does not approach, say 75° Oe, you will need to 'Chaptalise' (add sugar to the must). The Hydrometer Table should be used to ascertain the quantity of household sugar which is required. For example if the mustweight is 65° Oe, which is equivalent to 875g in 5 litres (28oz/gal) and you wish it to be 75° Oe (for 10% alcohol) then you must ADD 135g/5 litres (8oz/gal).

Now is the time to add the 'starter' and it is a good time to check and record acidity, but not to worry about it.

Wine should be as 'natural' as is possible and so even if an acid level is high (say 12g/l), it is not necessary to reduce it at this stage... The vintage should continue to be checked later and again before the last racking. Acidity **will** lessen and usually become satisfactory... and there is the possibility of 'spritzig' (A tiny prickle of Carbon Dioxide), which comes with a fortunate, but not certain, secondary fermentation by malo-lactic bacteria. (Appendix. Acidity).

Acid is a wine's spine. A wine which is insufficiently acid will be, at best, flabby and will fail to age with elegance. In the extreme, it will take on the taste of old fashioned cough mixture. It is always better to err towards a little too much acidity... which will fade with age and with advantage, than too little, which can do nothing but ill; and some modern yeasts, for example Gervin Strain 71B, will convert a proportion of malic acid into alcohol. A wine's bouquet and subtle tastes are due to terpenes and amino acids in the grape, and esters which are created mostly after the first fermentation and result mainly, from reactions of malic and succinic acids with the Ethyl and other alcohols in the wine. Also it is with this in mind, that the ferment is kept temperate, lest the bouquet evaporates.

At first the ferment will be fierce and often creating a large froth and so it is best to two-thirds fill the fermentation vessels and then to top right up as the fermentation subsides. Modern pure strain yeasts create far less froth than do others. The wine must very nearly fill the vessel as it stills and must always be under a fermentation lock. Fermentation will be over and the yeast will begin to fall after 7 to 14 days and it is then that a good yeast shows another of its qualities, for it will fall to a firm lees and quickly. 'Baker's' yeast and others, inferior for wine making, will settle neither fast nor firmly and will rise easily during racking.

Making Red Wine

When making Red Wine, (usually dark pink in the U.K.!) there is a choice. Either the pulp can be pressed within a day and the juice moved to a fermentation vessel, or, and in my own opinion with better results, it can be fermented 'on the pulp' and pressed 5 to 10 days later. This method maximises the colour and fruit extraction. Gervin FF is a particularly good yeast to use for this 'carbonic maceration', for it contains a terpene releasing enzyme and flavour enhancing terpenes are held, mainly, in the skin.

Beaujolais and Rioja are fermented using this 'Macération carbonique' method. Some growers de-stem, others do not... my choice depends upon whether the stems are soft or woody. In either event the grapes are piled into large epoxy-coated concrete vats together with a starter. The weight of the upper grapes crushes the lower and as the ferment moves through the mass, the grapes burst. The must is circulated and the grapes are kept submerged throughout the fermentation.

In scaling down the maceration method method to suit my needs, I have found that because I have never had a sufficient mass of grapes for efficient fermentation from weight alone, I first mill and then add the starter, cover and weight the pulp and turn it over morning and evening – for 5 to 8 days according to how the fermentation proceeds. Sufficient juice is taken at milling to check the Oechsle level.

When the pulp has been pressed the must is kept under a lock. Gervin FF continues working on the wine for some time after it falls and the 'feel' of the wine is greatly enhanced if it is not racked for some weeks. A few twists of the vessel each day will agitate the precipitated yeast into the must. The sediment is left unagitated for a few days before it is racked after four or five weeks.

Red wines will generally work to dryness and should be left so.

Racking Wines

Red and white wines **must** be taken off the lees within six weeks of standing or they will become 'mousy'. This is because the yeast's enzymes, which changed the sugars into alcohol and Carbon Dioxide, now turn their attentions to digesting the dead yeast.

The wine is racked through a wine syphoning tube, taking care not to take lees over with it.

This process must be done taking great care not to stir the lees... a good firm lees is a great help.

It is best to put some Carbon Dioxide into the bottom of the receiving vessel but, in any case the wine should enter low and the syphon end kept below the surface. Never splash it into the vessel. At this time crush and add one half of a Campden tablet per 5 litres (1 gallon) of wine. A further crushed one half of a tablet is added after each further racking.

It is possible to stop the ferment at any time by using a yeast excluding filter or, more brutally, by cooling and over-sulphiting. The latter is not good practice. The easiest and most usual practice is to work the wine to dryness and if desired, to sweeten it to taste at bottling. There is, however, absolutely no way in which a cane or beet sugared wine can provide the beautiful taste experience which comes from a wine fermented from nobly rotted grape extract of 120° Oechsle or higher... so sweet with natural sugars that some remains... Ways of approaching such delight are dealt with later.

Whenever there is further precipitation the wine must be racked and a further one half of a Campden tablet added. This practice keeps a level of around 50 parts per million of Sulphur Dioxide and will maintain must and bottle health. Sulphur Dioxide test kits may be obtained from homewine shops, but I have found that the described practice is sufficiently accurate.

The beneficial effects of sulphiting are many...

It prevents spoilage through infection.

Sulphur Dioxide is a reducing agent and can mop up any small amounts of Oxygen.

In the finished wine it prevents the oxidation of alcohols.

It exists in the wine as 'bound' and 'free'.

It binds onto Acetaldehyde, which is a link between sugar and alcohol and so it helps to prevent fermentation in the bottle and any reaction then is towards making glycerol. (There is up to 2% glycerol in fine 'Noble Rot' wines).

As a free agent it neutralises the minute electrical charges which suspend hazes and so helps to clear a wine.

The expectation is for four rackings. The final racking differs in that the wine is first fined and polished.

At this time you may see crystals adhering to the vessel's sides. These are Sodium and Potassium Tartrate crystals which are insoluble and tasteless. A sign of quality. It is yet another de-acidification and if it has not happened by this time then it may do so in the bottle. It does not happen in a flabby wine.

Acidity Check

After any Malolactic fermentation and before the last racking is the time to make a final check on acidity, if you believe it to be necessary. The most noble Riesling wines may have more than 9g/litre but the Riesling alone can achieve proper balance at that level and then only having been grown on Mosel slate. If your wine has more than 7.5g/litre acids then... wait and see... or if you feel it to be absolutely necessary, then reduce the acidity with precipitated chalk. Use 3 grams of chalk per 5 litres of wine for each gram of acid per litre greater than that which is required (Appendix). The reaction is violent... use a little wine in a large vessel. Add wine until the reaction ceases and then add to the bulk. This sudden neutralisation is a most unnatural and upsetting event for the wine and it will take it at least a fortnight to recover from this experience. It is best avoided. There will be a precipitate of, mostly, Calcium Malate with some Tartrate. Reduction of acidity will not be necessary if the grapes have ripened properly and all but a great excess will disappear during cellaring, and there is always the escape route of blending with a lower acid vintage.

My feeling is that the less you 'interfere' the happier is the wine. If you must intervene, then the intervention should be out of experience.

Fining and Polishing

It remains the practice in France with Red wines, to add white of egg whisked in with some of the wine in order to 'fine' and 'polish'.

A most convenient way is with a two stage fining, such as 'Kwikclear' or 'Gervin Two Stage' and half a Campden tablet. A little of the wine is mixed with the first stage and added to the bulk. After the recommended time, the second stage is mixed and added and the wine set aside for a few days, when it will have been cleaned and polished star bright and made ready for bottling. I used to filter after polishing but I believe now that, as German winemakers say, I 'stripped the coat from him', for each time a wine is fined or filtered it loses, not only unwanted suspension, but also degrees of bouquet and taste and colour.

Remember that wine is always open to infection or oxidation whenever it is without its fermentation trap.

Between the third and fourth racking and most likely during late March you may see thin streams of tiny bubbles rising from within the wine – not from the sediment. This is most likely to be caused by Malolactic bacteria which convert Malic acid into the less sharp Lactic acid (Milk acid). These bacteria are responsible for spoiling pickles and olives and flabby wines but they are beneficial in properly acid wines, not only in attenuating excess malic acid but in creating a refreshing 'prickle'.

Just as in 'real' Champagne or a good 'Moselle', some of the nascent Carbon Dioxide binds loosely with alcohol molecules, to be released in the glass and on the tongue.

Fining is best not done whilst the malolactic reaction is taking place, but this 'spritzig' should be captured by fining and either bottling or corking down into demijohns soon after it has completed.

Problems During Fermentation and Racking

If a must seems tardy in clearing put it outside for a few nights. Wine will freeze at around −4°C, but frosts which are deep enough to freeze a bulk of wine overnight are likely to have been forecast. This treatment also accelerates the crystallisation of Sodium and Potassium Tartrates.

Protein haze is colloidal and is usually difficult to treat and will not filter out. Protein stability tests are available and the treatment depends upon whether the haze is positively or negatively charged. It is a problem best avoided by having used Pectolase on milling.

Some winemakers use Bentonite – a diatomous earth – as an insurance against hazes, but it evolves relatively huge amounts of sediment and simply replaces one problem with another.

Heavy metal 'casses' should never happen, for there is no cause for juice or wine to contact Iron or Copper or any other heavy metal and the possibility for excess copper reaching the wine from spray is obviated by confining any copper spraying to the Winter.

Oxidisation – 'Madeirisation' is fine in Madeira or Marsala but it is the ruination of unfortified wines. Wine can oxidise only if it is left, untrapped, uncorked or undersulphited or if it lacks acid.

The same may be said for bacterial and wild yeast spoilage.

Bottling

Do not hurry to bottle. At worst the bottles may burst or your wine becomes fizzy and with an unsatisfyingly unfinished taste. Late April is quite soon enough to bottle white wines and even then it may be preferred to cork down in demijohns to bottle later.

Not necessarily before corking down in demijohns, but before bottling, is the time to decide sweetness.

Sweetening

Dry white is good, especially with meat or cheese and so is medium-dry, especially at picnics or 'just sitting in the sun'. Medium-sweet is around the sweetness of most of the poorer German wines to be found in supermarkets.

Wine has worked to 'dryness' when the gravity is below 1.000 (0 Oe). Usually around 0.97 (097 Oe).

My own white wine grapes are German and make pleasant varietal or blended wines which are sweetened or not for family tastes. German Wine Law classifies sweetness as follows…

Trocken (dry) ....less than 9g of sugars/litre

Halbtrocken (half dry or medium) ....9 to 18g of sugars/litre

Lieblich ('lovely') ....more than 18g of sugars/litre

A wine with 35g/litre of sugars or more is very sweet and most difficult to balance with acidity and alcohol. The very best way to attempt to emulate Trocken-Beeren-Auslese, Beeren-Auslese, Auslese, Tokaji or fine Sauternes is to attempt to do so, only, if the grapes have the potential for making a dessert wine naturally.

Dessert Wines

If there has been a superlative Spring and Summer the grapes will show that they are ripening well. As the sugars rise above 60° Oe the 'Noble rot' is welcome. You may encourage it by giving the vines polythene skirts from about the middle of September. These create good conditions for botrytis to develop and the vines benefit from the trapped sunshine by day and by night capture re-radiated heat.

Now, you hope for an excellent Autumn. You may be able to wait until December before picking... you decide just how much risk you are prepared to take. 100° Oechsle is as low a starting gravity which may result in a good 'Pudding wine' and even at this gravity the wine can work to dryness and require sweetening before bottling. An Oechsle level of 120° will generally cease working to leave around 20 degrees of sugars and 13-14 percent alcohol. It is possible to arrest fermentation, by chilling the must, racking and stabilising if you believe that things are going awry, but, the general rule is to interfere as little as you can.

Even if you fail to make the definitive English dessert wine you will have made a quite splendid table wine... and there is always next year. Siegerrebe, Bacchus and Schönburger have the potential for reaching very high must weights outside and Silvaner and Sauvignon Blanc can make superlative wines when absolutely ripe and with noble botrytis.

Sparkling Wines

The very cheapest sparkling wines are made by impregnating still table wines with Carbon Dioxide and if you have a pressure barrel such may be accomplished in just the same way as for home beer. But, just as artificially carbonated beer is inferior to beer carbonated by fermentation, so is impregnated wine a poor impersonation when compared with 'the real thing'. The quality dimension of the fermented sparkle is that a proportion of the nascent Carbon Dioxide molecules bond loosely with alcohol, to be released only upon contact with the sides of the glass and on the tongue. This is the 'spritzig' of good German wine and the lasting fizz of a good sparkling wine.

The phenomenon is accentuated when fermentation has taken place under pressure in a bottle, as has Champagne or Vintage Sekt. The essential requirements in still wine for it to become a palatable sparkling wine are that it should have at least 5 grams of acid per litre and should begin the bottle fermentation with 17/18 grams of sugar per litre. Usually it will have fermented out to be dosed with sugar syrup. The pressure in the bottle may reach more than 3 times atmospheric pressure and so only thick walled bottles of 'Champagne' style are to be used. The making of sparkling wine is fraught with danger... use ONLY sound bottles. It is best to add a measured dose of syrup to the bulk and then a starter 'Champagne' style yeast. For example Gervin Yellow label, which is a Pasteur Institute strain of **S. cerevisae bayanus** is most suitable for bottle fermentation.

Bottle immediately, filling to about 2" (5cm) below a hollow plastic stopper which is then secured by a wire muselet.

The bottles go onto their sides for a few days at 15-20°C (60-70°F) so that fermentation may begin. After a week or ten days 12°C is a high enough temperature to ensure that fermentation continues and the wine should be left alone for about a year save for a little gentle shaking once a month to redistribute the yeast. Always wear goggles and gloves and a hard hat when dealing with bottle fermenting wine.

After a year comes remuage – the gathering of the yeast prior to disgorging.

The bottles should be laid, neck downwards, on a rack canted at about 45 degrees. Every 3 days twist each bottle a little to the left and then, a little more sharply, to the right, so that the bottle finishes about 45 degrees further round at each visit. Eventually the yeast sediment will be gathered in the hollow stopper below a clear wine and it is ready to be disgorged.

By far the most convenient way to disgorge is by using a freezing mixture made by crushing ice with salt in the ratio of three to one. The efficiency of the mixture is very dependant upon an intimate admixture of small particles. The bottles, previously cooled to 5-7°C (40-55°F), are carefully planted 2-3" (50-75mm) into the mixture for about 20 minutes. Carefully – for the stopper and some of the wine will be at around –20°C (–8°F) – wipe the mixture from the neck and with the bottle pointing away from you at about 60 degrees and carefully remove the muselet and the hollow stopper. Top up the bottle with dry wine or syrup and reseal with a solid plastic stopper and muselet.

If dry wine is used the sparkling wine will be 'brut'. Other degrees of sweetness result from dosing the bottles with syrup.

'Extra dry' has 1.5-3g/litre sugar, 'Dry' has 2-4g/litre, 'Medium dry' up to 8g/litre and 'Sweet' up to 10g/litre. If the wine is sweetened it is essential that it is also dosed with stabiliser and sulphite.

The restoppered wine must be allowed to recover from the assault before drinking.

Finishing and Bottling

Red wines very often fall bright without assistance and are best left in demijohns for up to a year before bottling to allow the tannins to react and soften.

Now, whether it is because Coopering is a dying skill, or because of the 'Market', I do not know, but many of today's wine makers add 'Oakies' to the racked wine instead of barrel ageing. Oak chips in weighted muslin bags are sunk into the stainless steel storage tanks and, it is argued, act in the same way as oak barrels and soften the tannins. This cannot be strictly true since an oak barrel breathes and air has a part to play in barrel maturation. Nevertheless the oakies do change the wine in some way to lend a vanilla overtone to the wine. Oaking is, all too often overdone – and particularly by some New World and Antipodean Winemakers when it gives the wine a most intrusive taste of ice-cream.

If you like oaked wines and have no barrel, then try oakies, but… take care!

Dry wines can be bottled without much fear of refermentation.

There will be some live yeast around after other than the finest of filtering and it must be prevented from becoming activated by any added sugar. The securest way is to add Sodium Benzoate, which is a commonly used food preservative, and Campden tablets to the recommended dosage, at the time of adding sugar and immediately before bottling, but if you do not wish to use any additives, then the best way is to store the wine in demijohns and to bottle, to taste, half a dozen bottles at a time. This is best done a day or so before drinking to allow the wine to recover from the 'bottle sickness' which often follows bottling.

Do buy good quality corks and sit them in hot (not boiling) water, to which has been added a Campden tablet, for about half an hour, They will need weighting with a saucer.

You still need a corker… the French lever action type is best and obviates the necessity for 'flogging', for it flogs as it is loaded.

Using syphon tubing fill each bottle to within a cork length and a quarter of the top and cork immediately.

The appearance will be enhanced by encapsulation and by an informative label.

Storing

Chainstores and supermarkets have given an excellent service in introducing to our palates, fair to good wines from the world over, at prices which make most of them 'worth a try', but it is not in a wine's best interests to be stored upright, on a warm shelf and in a brilliant light! Keep yours on its side, in a stable temperature – a cellar is not a necessity – and in not more than subdued light… under the stairs perhaps.

Wine is alive and as it lives its life, it changes. Some wines reach and pass their prime quite soon after bottling… "Drink within six months of purchase". Some

Classic wines live for many years before they peak and may remain on a plateau of fineness for many more, before they decline and die.

United Kingdom wines are said to be at their best from one or two years after bottling, but I think that rule of thumb refers to such as Müller-Thurgau and Seyval, for maturation and ageing potentials differ with variety and treatment. What is certain, is that wine needs the backbone of good acidity for elegant maturation and longevity.

It is unlikely that your first wine will see two Christmases, but, after trying what is left after bottling and a perhaps a bottle in June, be patient until at least one year after the day of picking. The wine will have improved month on month.

Conclusion

It is too much to expect from our climate that a British Isles grower has fine enough fruit, from which to make great wines, but there are good, becoming better, wines being made here – particularly from the newer German varieties.

A fine grape variety carries a potential for fine wine. It rests with the winegrower to help the grape approach the possibilities allowed by its provenance and it is for the winemaker to help the ferment by intervening as little as possible and then, usefully. Experience comes from both good and bad vintages; from 'wait and see' but never from panic or brutality!

Keep records of all which you do and what comes about and act from experience.

That is the practice of the finest winemasters. Even after twenty five generations and more of working the same vineyard they continue the search for perfection.

It is certainly within the capability of any good gardener to produce better and more wholesome dessert grapes than he or she can expect to buy and to make individualistic wines, which have the potential to be as good as, and better than, many which they have bought.

Grape Juice

Grape juice will keep for at least a year if it is first pasteurised at 190°F (90°C) for two minutes. 'Grolsch' type or screw top bottles serve the purpose well. Add Vitamin C tablets (antioxidant) to each bottle at 250-350mg/litre before filling. The pasteurising temperature and the return to room temperature should be achieved as quickly as is sensibly possible. Store in the wine cellar.

Verjuice

Whilst it is unlikely that the grower's primary intention will be to produce verjuice, this alternative to vinegar may be made from bunches which have failed to ripen in the greenhouse or outdoors.

These are milled and pressed as for wine the 'verjuice' pasteurised and 'Grolsch' bottled.

Vinegar

'Vin aigre'... 'sour wine', is dilute acetic acid and is produced when one of some twenty or so species of **Acetobacter** digest alcohol in the presence of Oxygen.

The fruit or vinegar fly **Drosophila**, which can quickly infest a less than clean winery, cannot itself make vinegar, but is almost invariably a carrier of the bacteria.

Spirit vinegar, which is produced from commercial alcohol has only sourness, but wine, malt and cyder vinegars have the character of the malt or fruit.

Vinegar can be made from completely dry ferments of wine, beer or cyder.

Firstly the liquor should be diluted to contain less than 6% alcohol.

Half fill a container with the diluted liquor and plug the neck with cotton wool to allow free access of Oxygen but to exclude insects after inoculating the liquor with a good natural vinegar at the rate of 1:5 and maintain a temperature of 21-30°C (70–85°F).

After a short time the pellicle – **mycoderma aceti** (The Mother of Vinegar) – will, grow on the surface and great care must be taken to see that it is not fractured.

Acetification will take about three months – more if the temperature is lower – when the acidity will be about 4g/litre of acetic acid.

If the vinegar is racked off carefully, then more liquor can be added below the pellicle, taking care to keep it intact.

Freshly made vinegar is very harsh but improves with keeping to compare favourably with that on sale.

Bottles must be filled to the cap to reduce the chance of infection with **Acetobacter rancens**. The vinegar of commerce is usually pasteurised.

In Nature this bacterium completes a cycle… Water and Carbon Dioxide and sunshine… becomes firstly malic and then tartaric acid… which with warmth becomes the fruit sugars fructose and glucose… then, by fermentation, alcohol… then through the digestion by certain **Acetobacter** to vinegar… and finally, by the action of **Acebacter rancens**, it returns to water and Carbon Dioxide.

It is a sobering thought that the most sublime of wines, the most skilled of winemasters and the most appreciative of connoisseurs, is each no more than a brief hiatus in the endless loop of birth maturation and putrefaction.

Carpe diem!

… Show me which wine you drink and I will tell you which bottle you are…
German proverb

Chapter Six Summary

- ★ Ripening… the conversion of acids to sugars… requires a mean daily temperature of 10°C (50°F).
- ★ Dessert grapes finish with a higher sugar/acid ratio than wine grapes.
- ★ Protect the bloom on dessert grapes and cut with a "T".
- ★ Leave winegrapes until as fully ready as Autumn and courage allow.
- ★ Use no metal equipment other than aluminium or stainless steel for winemaking.
- ★ Use only 'Gervin' or other high quality and appropriate yeasts.
- ★ Do not panic over acidity, for it will attenuate… in extremis, back blend.
- ★ Rack often, fine once and filter once if you wish.
- ★ Keep records and learn from them.
- ★ Do not hurry to bottle or to drink.
- ★ The risk of bottles bursting is much reduced by working to dryness, storing in demijohns and adding sugar to taste immediately before bottling.
- ★ Store bottled wine on its side in a steady environment.

Appendix One

MONTHLY PLANNER FOR THE VINEYARD, VINERY AND WINERY

December

Finish any outstanding pruning and training outside and plant last cuttings. Keep all the vinery ventilators open. Spur prune older wood, lay the rods and spray with Tar Oil Winter wash. Put down organic matter. Rack and chill wine as necessary.

January

Dig over new ground. Keep the ventilators open. Rack and chill as necessary. Deal with bench grafting.

February

Carry out maintenance in vineyard. Keep the ventilators open. Rack as necessary and check that all containers are filled and trapped.

March

Plant out replacements or new vines. Keep the ventilators open and make sure that the vinery is in good repair for the coming season. Close down at night and give heat for Vinous varieties. Rack as necessary.

April

Planting should be finished. If it is sufficiently warm then some outdoor grafting can be done.

Lift and tie rods in the vinery as the sap reaches the tops. When budburst is imminent damp down regularly. Keep the ventilators closed at night and on cold sunless days, but no more heating is necessary. When certain, select good and alternate fruiting shoots about one foot apart and pinch out the others. Begin spraying programme a few days before flowering. This is a likely time to be bottling stable or stabilised wines.

May

Budbreak for outdoor vines.

After budbreak, pinch out non-fruiting shoots leaving a potential crop which the vine can carry without stress.

Full air on vinery unless frosty.

June

Keep next year's replacement canes in mind. Tuck in or tie fruiting shoots and hoe. Full flowering in the vinery. Keep on top of pinching out useless growth.

Keep vinery buoyant. Recommence spray programme after flowerset.

July

Vines will already or soon be in flower outside. Replacement canes are growing fast and should be attended to before pinching out fruiting laterals 6-8 leaves after the last bunch and keeping on top of pinching out sub-laterals and any non-fruiting laterals.

In the vinery begin thinning the bunches. Maintain spray regime. Keep removing useless growth.

August

Replacement canes should be long enough and stopped for the wood to ripen. Maintain spraying regime.

Attend to removing useless growth in the vinery and watch for veraison when all feeding and watering should cease.

The earliest varieties will ripen.

September

Outside early grapes to be netted in an early year. Remember you need to stop spray programme a month before picking.

Shut down the vinery each night to conserve heat for ripening. Watch for and deal with wasps. Begin heat for late Vinous varieties.

October

Outdoor grapes may look ripe, but be sure.

In the vinery the earlier varieties are nearly all ripe. Keep air on, on good days.

Late October may see the beginning of wine making. In any event check and clean equipment.

November

Vintage begins or continues. Later vintages may be blended back as wished. Pruning and training may begin as soon after harvest as is desired, or left as long as January.

Cease heating vinery after harvest. Suitable bunches can go to the grape cellar.

NOTE: The Calendar suggests actions but the timings will vary, from place to place, with variety and with timing of the seasons.

... He sleepeth well who wisely drinks.
Who sleepeth well no evil thinks.
Who thinks no evil never sins.
Who sinneth not, salvation wins.
Therefore he who drinketh well
Is certain to be saved from Hell.

(Medieval Monkish Doggerel)

APPENDIX TWO

TYPICAL ANALYSES OF SOME COMMON FERTILISERS

	N%	P%	K%
Hoof and Horn	13.0		
Superphosphate		16.0	
Sulphate and Potash			48.0
John Innes Base	5.1	6.4	9.7
National Growmore	7.0	7.0	7.0
Epsom Salt/Kieserite			20% Magnesium
Animal Manure/Compost	0.5	0.5	0.5
	With Humus and a variety of trace elements.		
Maxicrop Seaweed Extract			
'Tomato'	5.1	5.1	6.7
'General'	5.0	5.0	5.0
'Triple'	3.0	1.6	3.0

Maxicrop fertilisers/stimulant contain the trace elements...

Iron 165, Copper 83, Manganese 83, Zinc 29, Boron 43, Molybdenum 1 (mg/kg) (Soluble seaweed solids 16%), naturally occurring soil and systemic bio-stimulants.

	N%	P%	K%
Russian Comfrey var. 'Bocking'			
(Henry Doubleday, Ryton)		1.0	7.0

and... Calcium 2%, Iron 0.5%, Manganese 150 and Cobalt 1 (mg/kg).

WINE AND GRAPE EXTRACT... TYPICAL ANALYSES

Wine

Component	Amount	Effect
Water	80-93%	The carrier
Ethyl Alcohol	7-20%	Toxic but pleasing preservative
Glycerine	1-2%	The smoother and mellower
Carbon Dioxide	0.05-6.2%	Adds 'spritzig' and 'life'
Acetaldehyde	up to 40mg/l	Essential in alcohol process
Other Aldehydes	Traces	Bouquet
Esters	40-180mg/l	Bouquet
Higher Alcohols	up to 0.2%	Taste and hangover
Acetic Acid	up to 0.1%	Taste. Too much... ruination
Tartaric Acid, Malic, Succinic and Lactic	up to 0.8%	Give zing and esters
Other Acids	Traces	Flavour, bouquet
Unfermented Sugars		Dependent upon 'dryness'
Vitamin B Complex		Variable... remaining from yeasts
Tannins...		
White Wine	0.01 to 0.03%	Preservative and gives 'bite' and character
Red Wine	0.1 to 0.4%	Preservative and gives 'bite' and character

Grape Extract Typical Analysis of Mineral Content

(Parts/Million)

As Sulphates...

Potassium	0.001	Magnesium	8	Iron	0.6	Manganese	0.3
Zinc	0.3	Copper	0.08	Cobalt	0.001		

Other...

Molybdenum	0.001	Phosphorus	17	Calcium	7
Iodine	0.001	Fluorine	0.02		

Extract from... "Gesund mit Wein" (Heinz von Opel, 1988)

WINE VARIETIES... YIELD AND QUALITY POTENTIALS

NOTE: Some modern German varieties perform better in the U.K.

	Yield	Quality
Classic Varieties		
Cabernet Sauvignon	Low	Top
Chardonnay	Moderate but Consistent	Top
Chenin Blanc	Moderate to High	Variable
Merlot	Moderate to High	Top
Pinot Noir	Low yield essential for	Top
Riesling	Variable	Unsurpassable
Sauvignon Blanc	Low to Moderate	Moderate
Major Varieties		
Cabernet Franc	Moderate	High
Gewürtztraminer	Low to Moderate	Good
Gamay	Quite Productive	Medium/Good
Müller-Thurgau	Very High	Ordinary
Pinot Gris	Very Good	Medium
Other Varieties		
Auxerois	Moderate/High	Low/Medium
Bacchus	Good	Blowsy
Chasselas	Quite Productive	Insipid
Dornfelder	High	Promising
Ehrenfelser	Higher than Riesling	Good/Very Good
Faber	Lower than M-T	Better than M-T
Kerner	Higher than Riesling	Very Good

	Yield	Quality
Other Varieties (continued)		
Madeleine Angevine	Low if Cold	Respectable 'Muscat'
Ortega	Good	Rich/Flabby
Perle	Good	Ordinary/Floral
Regner	Reliable	Variable
Reichensteiner	Variable	Neutral
Scheurebe	Higher than Riesling	Very Good
Schönburger	Medium	Slightly Flabby
Siegerrebe	Low/Average	Very Rich
Würzer	High. Lowered for	Good

Abbreviated from 'Vines Grapes and Wines' (J. Robinson, 1986)

APPENDIX FIVE

CLIMATE COMPARISON... BRITISH ISLES/CONTINENTAL

Locality	Rainfall (mm)	*Heat as Degree Days
Norwich	382	955
Trier (Mosel)	399	1050

Extrapolated from European Weather Data 1960-1990.

*Degree days calculated from a datum of 10 degrees Celsius and from 1st May to 31st October inclusive.

Average Rainfall (mm) and Heat (degree Days (>10°C) 1/5-31/10) 1990-1996

Locality	Rainfall (mm)	*Heat as Degree Days
Brockford (Suffolk)	370	959
Trier (Mosel)	359	Greater than 1000

Method for computing Degree C. days...

1. (Total Maxima plus Total Minima) divided by twice the number of days in the month.
2. Subtract 10.
3. Multiply by number of days in the month.
4. Sum the Monthly Totals May to October inclusive.

HYDROMETER TABLE AND USE

About 47.5% of sugars are converted into alcohol, the rest becoming Carbon Dioxide, other alcohols, the growing yeast and energy.

Example Chaptalisation

The must weight is 65 degrees Oechsle which has the possibility for 8.6% alcohol. 75 degrees Oechsle is required weight for the wine to contain 10% alc… Each gallon of must will require 35.5 ounces Sugar ADDED TO – 27 3/4 ounces IN = 7 3/4 ounces. (1110g – 875g = 235g)

Specific Gravity	Sucrose in		Sucrose added to		Possible Alcohol	Gravity Oechsle
	1 gal	5 litres	1 gal	5 litres		
1.005	2 3/4 oz	85g	2 3/4 oz	85g		5
1.010	4 3/4	150	4 3/4	150	0.4	10
1.015	7	220	7 1/4	225	1.2	15
1.020	9	285	9 1/4	290	2.0	20
1.025	11	350	11 1/2	360	2.8	25
1.030	13 1/4	415	13 3/4	430	3.6	30
1.035	15 1/2	485	16	500	4.3	35
1.040	17 1/2	550	18	560	5.1	40
1.045	19 1/2	615	20 1/4	630	5.8	45
1.050	21 1/2	680	22 3/4	710	6.5	50
1.055	23 3/4	745	25 1/4	785	7.2	55
1.060	25 3/4	810	27 3/4	865	7.9	60
1.065	27 3/4	875	30 1/4	945	8.6	65
1.070	30	945	33	1030	9.3	70
1.075	32	1010	35 1/2	1110	10.0	75
1.080	34 1/2	1075	38 1/2	1200	10.6	80
1.085	36 1/2	1140	41 1/4	1285	11.3	85
1.090	38 1/2	1205	44	1370	12.0	90
1.095	40 3/4	1275	47	1465	12.7	95
1.100	42 3/4	1340	49 3/4	1550	13.4	100
1.105	44 3/4	1405	53	1645	14.2	105
1.110	47	1475	56	1745	14.9	110
1.115	49	1540	59	1845	15.6	115
1.120	51 1/4	1605	63	1965	16.3	120
1.125	53 1/4	1675	67	2090	17.1	125
1.130	55 1/2	1740	71	2215	17.8	130

APPENDIX SEVEN

FERMENTATION PROCESS... MUCH SIMPLIFIED

Yeast **Saccharomyces elipsoideus** varieties use some half a dozen enzymes to digest sugars and in the following sequence...

1. Added Sucrose (household sugar), which has 12 Carbon atoms is inverted to 6-Carbon sugars similar to fruit sugars.
2. 6-Carbon Glucose and Fructose are converted by enzymes to 3-Carbon 'Triose' sugars.
3. The triose sugars are digested into Glyceric acid, and...

   ... an **End Product**........................**Glycerine**
4. Glyceric acid, is changed ... to ....Pyruvic acid
5. Pyruvic acid... acted upon by yet another enzyme becomes

   ... an End Product..............**Carbon Dioxide**

   ... and...
6. Acetaldehyde, which becomes ....Glyceric acid
7. Glyceric acid is either unchanged and loops to Stage 4 or... produces

   ... the **End Products** ....**Succinic acid**

   ... and...

   ....**Ethyl alcohol**

DIY PRESS

Press Frame. T-Iron sufficient for two 23" square frames. Welded and 4 coats Hammerite. Screwed to underside of underplate.

Screw old bus jack with 10" traverse.

Arm Angle Iron and pipe 4 coats Hammerite.

Screw collar and plate bolted under upper frame.

Base. All 4" x 2" oak 1/2" coach bolted or 1/2" threaded rodded.

Feet 19" 2 off.

Legs 16" two off.

Inner crosspieces 12" two off.

Pressplate supports 19" two off.

Underplate 18" x 12" x 3/4" ply screwed to

Pressplate 18" square 3/4" ply two off glued together.

Pressplate sides 17" x 21/2" x 1" hardwood four off, glued and screwed to pressplate. Pressplate drilled to take 3/4" **Exit Pipe**.

Basket 34" x 1" x 1/8", aluminium two off ... 1" lugs each end bent at right angles. 1/4" stainless bolts and nuts two each off. 13" x 3/4" x 1/2" oak thirty eight off, stainless screwed edgewise, inside aluminium straps.

Loose Basket base 11" square 3/4" ply routed one side.

Press piston 8" diameter 3/4" ply three off. Two glued together and one loose and routed one side.

Inner basket and underpiston from stout plastic windbreak.

Fruit bag from coarse nylon net.

All wood and metal which can be in contact with grape juice to be finished 4 coats Yacht varnish.

DIY GRAPE MILL

The illustration includes a separate base to which fits either a cider apple grinder or the illustrated grape mill. The only essentials for a mill are geared rollers. The following description is simply typical.

The rollers may be of any size... in this case two old 13" grasscutter rollers secured onto 1/2" threaded rod and bound with plastic garden line for grip and finished with four coats of Yacht varnish. **The frame** is of 3/4" outdoor ply glued and screwed. Four ex W.D. ball races are sunk half way in on the outside at centres appropriate for two ex-lathe gear-wheels. These gear wheels are of different sizes, to accomplish a tearing action and were selected from an assortment to allow the rollers to be 1/8" to 3/16" apart (5mm) and secured with nuts and locknuts. All metal parts which can contact the juices are finished with four coats of Yacht varnish. The gears are lubricated with Vaseline.

The hopper is of 1/4" outdoor ply and stock hardwood, finished with four coats of Yacht varnish.

The wheel is a Hayter cutter pulley and the handle is 1/2" pipe secured by a coach bolt. Again varnished four coats.

The intention was to use a small electric motor but this has yet to be accomplished.

MEASURING ACIDITY

Required... 25ml pipette; 50ml burette; 150ml conical flask; N/3 Sodium Hydroxide; 1% Phenolphthalein.

Method Assumption... all equipment cleaned with deionised water.

Fill burette, note level; Pipette exactly 25ml wine into flask. Add one drop Phenophthalein and titrate with N/3 solution a few drops at a time until pink colour will not fade on agitating; Note level.

Carry out the titration three times noting starting and finishing levels; Sum the differences and divide by three (if they are reasonably similar).

Result... each 1ml of reagent is Equivalent to 1gm/litre of Tartaric acid.

Watch acidity but do not hurry to correct. If above 8.8g/litre AFTER the third racking then reduce for each gram/litre excess acid.

Acidity kits are available from Home Wine Stores. Note that each manufacturer will include differing methods... READ THE INSTRUCTIONS.

APPENDIX ELEVEN

SUPPLIES... VINEYARD ASSOCIATIONS... BIBLIOGRAPHY

Vines

East	Reads, Hales Hall, Loddon, Norfolk
London	RHS Gardens, Wisley, Surrey
Midlands	Ashywood Nurseries, Ripon Road, Killinghall, North Yorks HG32 2AY
Scotland	Tweedies Fruit, Maryfield, Dumfries DG2 0TH
South	Deacons Nursery, Moorview, Godshill, Isle of Wight PO38 3HW
Wales/West	Sunnybank, Pontrilas, Herefordshire HR2 0BX
Wholesale only	D. Pritchard, Worten Courtney, Somerset Minimum order 10 vines.
Ornamental Vines	Hilliers, Winchester, Hampshire ...and also see **RHS PLANTFINDER**

Sprayers

Cobra Sprayers	Shepherd's Grove Industrial Estate, Suffolk

Winepresses, Mills, Yeasts, General Winemaking Chemicals and Equipment

Boots Chemists. Some stores carry more than others.
Harts Homebrews, 20 Bury Street, Stowmarket, Suffolk.
Homebrew and Wine Centre, 35 Crouch Street, Colchester, Essex.
Luton Winemakers Stores, 125 Park Street, LU1 3HG

Vineyard Associations

At the time of writing 'English and Welsh' Vineyards are undergoing change to 'U.K. Vineyards Association' which is an Association of individual Regional Associations.

Information regarding your local Association and Vineyards which welcome visits... contact... U.K. Vineyards Association, Bruisyard Vineyard, Framlingham, Suffolk. (01728 638281)

Bibliography

H. Baker & R. Waite	Grapes Indoor & Out	R.H.S.	1993
R. Croft-Cooke	Madeira	Putnam	1966
G. Ordish	The Great Wine Blight	Sidgwick Jackson	1987
G. Pearkes	Vinegrowing in Britain	J. M. Dent	1989
N. Poulter	Growing Vines	Amateur Winemaker	1977
J. Robinson	Vines, Grapes and Wines	Mitchell Beazley	1986
A. Toogood	Grapes	Collins	1992
B. Turner & R. Royeroft	Winemaker's Encyclopaedia	Faber & Faber	1979
P. Vandyke Price	Wine Lore, Legends, and Traditions	Hamlyn	1985
Pro Riesling	Der Riesling und Seine Weine	Wiesbadener Graphische	1986
H. von Opel	Gesund mit Wein	Woschek	1988

Appendix Twelve

VINEYARDS IN GERMANY KNOWN TO THE AUTHOR WHICH WELCOME U.K. VISITORS

Weindomäne/Weinmuseum Schlagkamp
56820 Senheim/Mosel.

Weingut Menger
Hauptstrasse 12
67575 Eich/Rheinhessen.

Weingut Schales
Aylzeyer Strasse 160
67592 Flörsheim-Dalsheim/Rheinhessen.

A complete list of German Vineyards welcoming visits may be obtained from...

German Wine Information Service,
Lane House,
24 Parson's Green Lane,
London
SW6 4HS

INDEX

Auxerois 17

Bacchus (rebe) 21
Birds 86
Black Hamburg 33, 58, 69
Blau Portuguese 28
Blaufrankisch 28
Botrytis 88, 98
Brant 29
Buckland Sweetwater ...34, 69

Cabernet Franc 25
Cabernet Sauvignon 25
Calabresser 23
California
- University of 8
- Mission 12

Campbell's Early 29
Campden tablet 97, 104, 105, 110
Cannon Hall Muscat 33
Carbon dioxide 75, 100-108, 119, 124
Chaouch 33
Chardonnay 17, 40
Chasselas 24, 33
- clones 17
- **de Fontainbleau** 10

Chile 8, 12
Chenin Blanc 17
Concord 29
Currant 11

Dead Arm (Phomopsis) 88
Deckrot 37
Dornfelder 28
Downy Mildew (Perenospera viticola) 87
Dunkelfelder 28, 36

Edelfaule 98
Entipiose virus 90

Faberrebe 21
Fan Leaf virus 90
Farbertraube 28
Feeding 74
Fendant 10, 17
Fertiliser 41
- Blood fish bone 80
- Boron 78
- Compost 76
- Growmore 76
- Iron 79
- J. I. Base 76
- J. I. 3 81
- Lime/Calcium 76
- Magnesium 77-78
- Manganese 78-79
- Maxicrop 78
- Nitrogen 76
- Phosphorus 77
- Potash 77
- Russian Comfrey 76
- Seaweed 80
- Shoddy 80
- Trace elements 80
- Zinc 78-79

Findling 21
Florida 7, 8
Fort Caroline 7
Frölich 28
Fructose 93, 124
Fruit
- budburst 47, 80, 88, 115
- set 63
- thinning 63-64
- veraison ...62, 65, 80, 93, 116

Gagarin Blue 32, 64, 97-98, 101
Gamay 23, 25, 52
Gewürztraminer 18, 23, 24
Glucose 93, 113
Grape
- Migration 9-12
- Juice 111

Grapes
- Coldhouse 32-34
- Hothouse 34-35
- Muscat 32
- Ornamental 36-37
- Outdoor 15-32
- Recommended cultivars 35, 39
- Sweetwater 32
- Vinous 32
- Wall 35, 39

Grey Mould (Botrytis) 88
Gutedel 10, 17

Harvesting
- general 93-98
- Winegrapes 96

Helfensteiner 28
Heroldrebe 28
Huguenots 7-8
Himrod Seedless 29

Jubilaumsrebe 28

Kanzler 21
Kerner/Kernling 21
Kuhlmann 27
Kuibishevski 32

Leaf Roll virus 90
Leon Millot 27
Lexia 10
Lugliencabianca 23

Madeira 10-11
- Columbus 12

Madeleine Angevine 18, 23, 24, 90
Madeleine Royale 18, 33
Madeleine Silvaner 18
Malic Acid 93, 97, 100, 102, 106
Malvasia 10-11
Mealy Bug 85, 87
Merlot 10, 25
Morio-Muscat 21
Mosaic virus 90
Mrs. Pince Black Muscat ...34
Mulching 74
Müller-Thurgau 16, 20, 21, 24, 90, 96
Müllerrebe 25, 28
Muscat Hamburg 24, 34
Muscat Hamburg/ Gros Colmar 34, 71
Muscat of Alexandria ...10, 33

No-Spray (Phoenix) 25
Noble Rot 98, 107

Oenology/Ampelography 9
Ontario 29
Ortega 23

Pais 12
Pectolase 101, 106
Perle 23
Perle de Czaba 18, 52
Perlette 18

Pests & diseases
spraying formulae 89
Petit Verdot 25
Phoenicians and Greeks 10
Phoenix 25
Phylloxera
Australia &
New Zealand 8
California 8
Chile 8
Europe 8
Mexico 8
South Africa 8
Suffolk 8
Sussex 8
vastatrix 12-13
Pierce's disease 7-8
Pineau de la Loire 17
Pinot Blanc 21
Pinot Gris 18
Pinot Meunier 25
Pinot Noir 10, 24, 27, 40, 54
Planting
distances 42
preparation 41
site selection 40
time 41
Pourriture Noble 98
Powdery Mildew
(Oidium Tuckeri) ...12, 13, 87
Précoce de Malingre
............18, 47, 52
Propagation
cuttings 70
grafting 71
layering 71
vine eyes 70

Red Spider Mite 83-85, 87
Regner 23
Reichensteiner 23, 48
Riesling 20-24, 105
'German' 23
Roem van Boskoop 28, 63
Roman Empire 11
Rootstocks 37-38
125AA 38
3309 38
5BB 38
5C 38
SO4 38
Rülander 18

Sauvignon Blanc 18, 108
Scale Insect 84
Scheurebe 24
Schiava 33
Schönburger 24, 108
Schuyler 29
Seibel 24, 27
5278 (Aurora) 24
Seyval 24
Siegerrebe 24, 40, 90
Silvaner 18, 20, 21, 24, 90
Sodium benzoate 110
Spätburgunder 28
St. Laurent 28
Sugars 93
Sulphur dioxide 104
Sultana 11
Summary
Chapter 1 14
Chapter 2 39
Chapter 3 73
Chapter 4 82
Chapter 5 92
Chapter 6 114
Syrah 10

Tartaric Acid 20, 93, 113
Tokay d'Alsace 18
Training
Alsatian 54
Bergerac 54
Chablis/Champagne 54
Geneva Double Curtain ...52
Goblet 52
Guyot Ancient varieties ...47
Guyot Double 43
Guyot fence 44
Guyot high 54
Guyot Single 45
Loire 54
Mosel Heartshaped 48
Pots 67-68, 81
Recommendations 55
The Curate's Vinery 66
The Trier Wheel 51
Under cover 57-62
Valée de la Marne 54
Victorian Table
Decoration 67-68
Victorian vinery 57
Wall 56
Trebbiano/Trebularum 20
Triomphe d'Alsace 27
Trollinger 21, 28, 33, 58

Ugni Blanc 20

Vitis 'Brant' 36
amurensis 36
berlandieri 7, 37-38
betulifolia 36
californica 36
coignetiae 36
Davidii & var. 36
hybrid... EU definition...15
labrusca 7
pulchra 36
qualities required 15-16
range/season/key 16-17
riparia 36, 37, 38
ripening comparisons 90
rotundifolia 7
rupestris 7, 37
sezzanensis 7
The vine 7
vinifera 7
ornamental 36
subspecies sylvestris 23
X V. amurensis 15
Valentina Tereschkova 32, 97
Verjuice 112
Vine Weevil 85-86
Vinegar 112-113
Vitamin C 111

Wash-Galway line 16
Wasps 86, 116
Whitefly 84
Wine
acidity 105
bottling 107
Civilisation 9
Deliberate
manufacture of 8
dessert 107
equipment 98
fermentation 100
fining/polishing 105
Liebfraumilch 20
milling/pressing 101
Myths and Stories 8-9
problems 106
Quality 15
racking 103
red 103
sparkling 108
storing 110
sweetness 107
Table 15, 27
Winter Wash 89
Wrotham Pinot 10, 25
Würzer 24

Zinfandel 29
Zweigeltrebe 28

LIST OF ILLUSTRATIONS IN COLOUR

A Small Garden Vineyard *61*
Bacchus (rebe) Cover *22*
Black Hamburg *30*
Campbell's Early *30*
Chaouch *31*
Chasselas Rose *31*
Coldhouse Vines *61*
Das Trierrad Training *49*
DIY Grape Mill 99
DIY Wine Press 99
Edelfaule *94*
Ehrenfelser *22*
Gagarin Blue *30*
Gewürtztraminer *19*
Gutedel *31*
Kerner *22*
Madeleine Royale *31*
Madresfield Court *2*
Mosel Training *49*
Mrs. Pince's Black Muscat *30, 31*
Müller-Thurgau *19*
Muscat Hamburg *16*
Noble Rot 94
Outdoor Vines, Pruned *61*
Pinot Gris *19*
Pot Vines
  One and Two-year old 68
  Two-year, pruned 68
Riesling *22*
Rivaner *19*
Rülander *19*
Scheurebe *26*
Schönburger *26*
Siegerrebe *26*
Silvaner *26*
Tereschkova *30*
Veraison *16*, 94
Winemaking Equipment 99

LIST OF LINE DRAWINGS

Grape Cellar Rack *95*
Victorian Grape Table Decoration *64*
Vine
  Cuttings *70*
  Grafting *71*
  Layering *71*
  Parts of *41*
  Pests *85*
Vine Training
  Alsatian *53*
  Bergerac *55*
  Casenave Spur *56*
  Champagne/Chablis *55*
  Curate's Vinery *66*
  Das Trierrad *51*
  Geneva Double Curtain *53*
  Glasshouse *58*
  Goblet *51*
  Guyot double *44*
  Guyot high *53*
  Guyot single *44*
  Loire *55*
  Mosel hertzliche *50*
  Pots *67*
  Umbrella *53*
  Valée de la Marne *55*
  Wall *56*